Skills Worksheet)

Directed Reading

Section: The Origins of Genetics

Read each question, and write your answer in the space provided.

1. What did T. A. Knight discover?

____

____

____

2. How did Mendel's scientific work differ from the work of T. A. Knight?

____

____

____

3. What are three reasons the garden pea is a good subject for studying heredity?

____

____

____

Complete each statement by writing the correct term or phrase in the space provided.

4. A mating that considers one pair of contrasting traits is called a(n)

_____________________________ cross.

5. The first two individuals that are crossed in a breeding experiment are called

the _____________________________ generation.

6. In Mendel's experiment, the _____________________________ generation was

obtained by cross-pollinating the P_1 generation.

7. The _____________________________ generation in Mendel's experiment showed
both forms of the trait in a ratio of 3:1.

Directed Reading

Section: Mendel's Theory

In the space provided, write the letter of the description that best matches the term or phrase.

_______ **1.** alleles

_______ **2.** dominant

_______ **3.** recessive

_______ **4.** homozygous

_______ **5.** heterozygous

_______ **6.** genotype

_______ **7.** phenotype

a. when two different alleles are present, the allele that is completely expressed

b. when two alleles of a particular gene are the same

c. when two alleles of a particular gene are different

d. an organism's physical appearance

e. the set of alleles that an organism has

f. different versions of a gene

g. when two different alleles are present, the allele that has no observable effect on the organism's appearance

Complete each statement by writing the correct term or phrase in the space provided.

8. If the allele for yellow peas is Y, the allele for the contrasting trait, green peas,

is _____________________ .

9. If Tt is the genotype of a plant, where T stands for tall and the recessive allele

stands for short, its phenotype is _____________________ .

10. If tt is the genotype of a plant, where T stands for tall and the recessive allele

stands for short, its phenotype is _____________________ .

Read each question, and write your answer in the space provided.

11. What is the law of segregation?

____

____

____

12. What is the law of independent assortment?

____

____

____

Directed Reading

Section: Studying Heredity

Complete each statement by writing the correct term or phrase in the space provided.

1. In a test cross to determine if an individual with a dominant phenotype is heterozygous or homozygous for the trait, you always cross the individual with a homozygous _________________________ individual.

2. If the offspring of a test cross all have the dominant trait, then the genotype of the individual being tested is _________________________ .

3. If some of the offspring of a test cross have the recessive trait, then the genotype of the individual being tested is _________________________ .

4. The probability that a gamete from a plant with a Tt genotype will carry a t allele is _________________________ .

5. The probability of homozygous recessive offspring resulting from a cross between two homozygous dominant individuals is _________________________ .

6. The probability of heterozygous offspring resulting from a cross between two heterozygous individuals is _________________________ .

Read each question, and write your answer in the space provided.

7. When studying a pedigree, how do scientists determine if a trait is sex-linked or autosomal?

____

____

8. When studying a pedigree, how do scientists determine if a trait is dominant or recessive?

____

____

____

____

Directed Reading

Section: Complex Patterns of Heredity

In the space provided, explain how the terms in each pair differ in meaning.

1. polygenic trait, multiple alleles

____

____

____

2. incomplete dominance, codominance

____

____

____

____

Complete each statement by writing the correct term or phrase in the space provided.

3. Sometimes genes are damaged or are copied incorrectly, resulting in faulty

______________________________ .

4. ______________________ ______________________________ is a genetic disorder caused by a defective gene that makes a protein necessary to pump chloride in and out of cells.

Read each question, and write your answer in the space provided.

5. Who should go for genetic counseling prior to having children?

____

____

____

6. What is gene therapy?

____

____

Active Reading

Section: The Origins of Genetics

Read the passage below. Then answer the questions that follow.

Mendel's initial experiments were monohybrid crosses. A **monohybrid cross** is a cross that involves one pair of contrasting traits. For example, crossing a plant with purple flowers and a plant with white flowers is a monohybrid cross. Mendel carried out his experiments in three steps.

Step 1: Mendel allowed each variety of garden pea plants to self-pollinate for several generations. This method ensured that each variety was **true-breeding** for a particular trait; that is, all the offspring would display only one form of a particular trait. For example, a true-breeding purple-flowering plant should produce only plants with purple flowers in subsequent generations.

These true-breeding plants served as the parental generation in Mendel's experiments. The parental generation, or **P generation,** are the first two individuals that are crossed in a breeding experiment.

Step 2: Mendel then cross-pollinated two P generation plants that had contrasting forms of a trait such as purple and white flowers. Mendel called the offspring of the P generation the first filial generation, or **F_1 generation.** He then examined each F_1 plant and recorded the number of F_1 plants expressing each trait.

Step 3: Finally, Mendel allowed the F_1 generation to self-pollinate. He called the offspring of the F_1 generation plants the second filial generation, or **F_2 generation.** Again, each F_2 plant was characterized and counted.

SKILL: READING EFFECTIVELY

Read each question, and write your answer in the space provided.

1. The prefix *mono-* means "one." How does this apply to the key term *monohybrid cross*?

____

____

2. What information does the third sentence tell the reader?

____

____

Active Reading *continued*

3. Describe the offspring of a true-breeding white-flowering plant.

____

____

4. What is the P generation?

____

____

5. What does the term F_1 *generation* refer to?

____

____

SKILL: INTERPRETING GRAPHICS

The figure below shows three generations of plants. Insert the following labels in the spaces provided: cross-pollination, F_1, F_2, P, self-pollination.

6. _________ generation   **7.** _________ generation   **8.** _________ generation

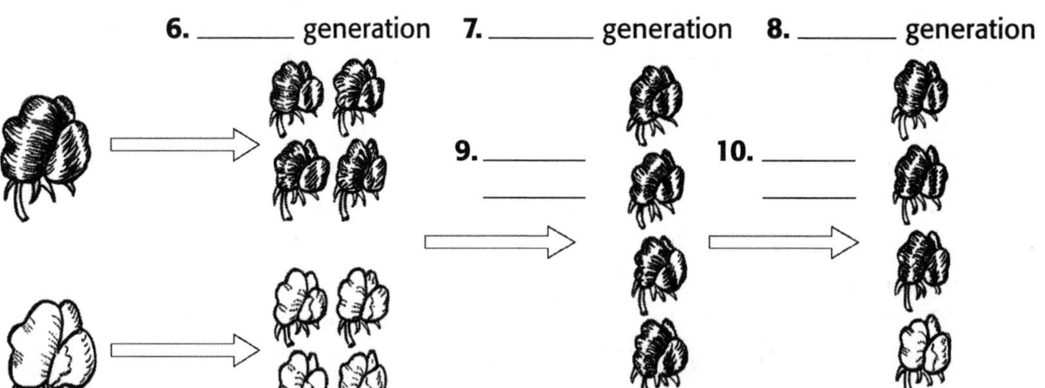

9. _________

10. _________

In the space provided, write the letter of the phrase that best completes the statement.

_______**11.** During the course of his experiment, Mendel studied traits in
   a. one generation of plants.
   b. two generations of plants.
   c. three generations of plants.
   d. more than five generations of plants.

Active Reading

Section: Mendel's Theory

Read the passage below. Then answer the questions that follow.

Geneticists have developed specific terms and ways of representing an individual's genetic makeup. Letters are often used to represent alleles. Dominant alleles are indicated by writing the first letter of the trait as a capital letter. Recessive alleles are also indicated by writing the first letter of the dominant trait, but the letter is lowercase.

If two alleles of a particular gene present in an individual are the same, the individual is said to be **homozygous** for that trait. If the alleles of a particular gene present in an individual are different, the individual is **heterozygous** for that trait.

SKILL: READING EFFECTIVELY

Read each question, and write your answer in the space provided.

1. How are dominant alleles often represented?

____

____

2. How are recessive alleles often represented?

____

____

3. A particular plant is said to be homozygous for seed color. What does this mean?

____

____

4. Another plant is said to be heterozygous for flower color. What does this mean?

____

____

Active Reading *continued*

5. The allele for yellow peas is dominant to the allele for green peas. How would you represent the alleles of a plant that is heterozygous for seed color?

____

____

6. The allele for purple flowers is dominant to the allele for white flowers. How would you represent the alleles of a plant that is homozygous recessive for flower color?

____

____

7. How would you represent the alleles of a plant that is heterozygous for flower color?

____

____

In the space provided, write the letter of the phrase that best completes the statement.

_______ **8.** A plant with YY alleles for seed color is
    a. heterozygous dominant for this trait.
    b. homozygous dominant for this trait.
    c. homozygous recessive for this trait.
    d. Either (a) or (b)

Skills Worksheet

Active Reading

Section: Studying Heredity

Read the passage below. Then answer the questions that follow.

A **Punnett square** is a diagram that predicts the expected outcome of a genetic cross by considering all possible combinations of gametes in the cross. Named for its inventor, Reginald Punnett, the Punnett square in its simplest form consists of four boxes inside a square. The possible gametes that one parent can produce are written along the top of the square. The possible gametes that the other parent can produce are written along the left side of the square. Each box inside the square is filled with two letters obtained by combining the allele along the top of the box with the allele along the side of the box. The letters in the boxes indicate the possible genotypes of the offspring.

SKILL: READING EFFECTIVELY

Read each question, and write your answer in the space provided.

1. What information does the first sentence tell the reader?

____

____

2. What do letters written along the top and left side of a Punnett square represent?

____

____

3. How is the combination of letters inside each square determined?

____

____

4. What do the letters in the boxes indicate?

____

____

5. What data did Mendel obtain when he examined each F_1 plant?

____

____

Active Reading *continued*

SKILL: ORGANIZING INFORMATION

The figure below shows a Punnett square. It shows a cross between two pea plants that are heterozygous for seed color. Use the Punnett square to answer the questions that follow. Write your answers in the spaces provided.

Yy
(Heterozygous)

	Box 1	*Yy*
Yy (Heterozygous)	*Yy*	Box 4

6. What pair of letters should appear in Box 1?

____

7. What pair of letters should appear in Box 4?

____

8. How many homozygous dominant offspring would be produced?

____

9. How many homozygous recessive offspring would be produced?

____

10. How many heterozygous offspring would be produced?

____

11. How many of the offspring would have green seeds?

____

12. How many of the offspring would have yellow seeds?

____

In the space provided, write the letter of the term or phrase that best completes the statement.

_______**13.** Each box inside a Punnett square represents one

    a. allele.   **c.** dominant trait.

    b. parent.   **d.** offspring.

Active Reading

Section: Complex Patterns of Heredity

Read the passage below. Then answer the questions that follow.

Genes with three or more alleles are said to have **multiple alleles.** When traits are controlled by genes with multiple alleles, an individual can have only two of the possible alleles for that gene. For example, in the human population, the ABO blood groups (blood types) are determined by three alleles, I^A, I^B, and i. The letters A and B refer to two carbohydrates on the surface of red blood cells. The i allele means that neither carbohydrate is present. The I^A and I^B alleles are both dominant over i, which is recessive. But neither I^A nor I^B is dominant over the other. When I^A and I^B are both present in the genotype, they are codominant.

SKILL: READING EFFECTIVELY

Read each question, and write your answer in the space provided.

1. What information does the first sentence convey to the reader?

____

____

2. Why does the term *blood types* appear in parentheses in the third sentence of this passage?

____

____

3. What do the letters A and B refer to in the alleles I^A and I^B?

____

____

4. What allele is dominant for human blood type? What allele is recessive for this trait?

____

____

5. What causes an individual to show both the I^A and I^B forms of the trait for human blood type?

____

____

Active Reading *continued*

In the space provided, write the letter of the phrase that best completes the statement.

_______ **6.** In humans, the *i* allele for blood type means that
     a. one kind of carbohydrate is on the surface of red blood cells.
     b. two kinds of carbohydrates are on the surface of red blood cells.
     c. more than three kinds of carbohydrates are on the surface of red blood cells.
     d. neither carbohydrate is present on the surface of red blood cells.

Vocabulary Review

In the space provided, write the letter of the description that best matches the term or phrase.

_______ 1. heredity

_______ 2. genetics

_______ 3. monohybrid cross

_______ 4. true-breeding

_______ 5. P generation

_______ 6. F_1 generation

_______ 7. F_2 generation

_______ 8. alleles

_______ 9. dominant

_______ 10. recessive

_______ 11. homozygous

_______ 12. heterozygous

_______ 13. genotype

_______ 14. phenotype

_______ 15. law of segregation

_______ 16. law of independent assortment

a. the alleles of a particular gene are different

b. the two alleles for a trait separate when gametes are formed

c. the alleles of different genes separate independently of one another during gamete formation

d. not expressed when the dominant form of the trait is present

e. passing of traits from parents to offspring

f. all the offspring display only one form of a particular trait

g. the expressed form of a trait

h. first two individuals crossed in a breeding experiment

i. physical appearance of a trait

j. a cross that considers one pair of contrasting traits

k. offspring of the F_1 generation

l. when the two alleles of a particular gene are the same

m. branch of biology that studies heredity

n. different versions of a gene

o. offspring of the P generation

p. set of alleles that an individual has

Write the correct term from the list below in the space next to its definition.

codominance	pedigree	Punnett square
incomplete dominance	polygenic trait	sex-linked trait
multiple alleles	probability	test cross

_______________________ **17.** diagram that predicts the outcomes of a genetic cross

_______________________ **18.** cross of a homozygous recessive individual with an individual with a dominant phenotype of unknown genotype

_______________________ **19.** the likelihood that a specific event will occur

_______________________ **20.** a family history that shows how a trait is inherited

_______________________ **21.** trait whose allele is located on the X chromosome

_______________________ **22.** when several genes influence a trait

_______________________ **23.** when an individual displays a trait that is intermediate between the two parents

_______________________ **24.** two dominant alleles are expressed at the same time

_______________________ **25.** genes with three or more alleles

Science Skills

Analyzing Experiments

Part of Gregor Mendel's hypotheses of inheritance proposed that when two contrasting traits occur together, one of them may be completely expressed, while the other may have no observable effect on the organism's appearance. He called the expressed trait dominant and the unexpressed trait recessive. In his genetic experiments, Mendel studied seven contrasting traits of peas. One of the traits he studied was the trait for seed shape. Mendel found that there are two forms of the trait for seed shape—wrinkled and round. In an experiment to determine which form of this trait was dominant, Mendel performed the following experiment:

A. Mendel allowed the plants to self-fertilize to produce the homozygous P generations.

B. Mendel then performed a cross between the homozygous plants with round seeds (RR) and homozygous plants with wrinkled seeds (rr) to produce the F_1 generation.

C. Mendel then allowed the F_1 plants to self-fertilize to produce the F_2 generation.

D. Mendel analyzed the results of the crosses to determine the genotypes and phenotypes present.

Use the information above to answer questions 1–5.

1. To determine the results of this experiment, complete the Punnett squares below by writing the correct allele(s) in the space provided.

P   *RR* × *rr*

F_1 a. _______   b. _______

c. _______   d. _______   e. _______

f. _______   g. _______   h. _______

F_2 i. _______   j. _______

k. _______   l. _______   m. _______

n. _______   o. _______   p. _______

Science Skills *continued*

Read each question, and write your answer in the space provided.

2. What phenotypes are present in the F_1 generation? What genotypes are present?

____

____

____

____

3. What phenotypes are present in the F_2 generation? In what ratio are they present?

____

____

____

____

4. Does your analysis support or refute Mendel's hypothesis of dominant and recessive inheritance? Explain.

____

____

____

____

5. Do you think the ratios of F_2 phenotypes that Mendel observed in his experiment were exactly the same as the F_2 phenotype ratio that you calculated? Explain.

____

____

____

____

Concept Mapping

Using the terms and phrases provided below, complete the concept map showing the principles of genetics.

codominance	multiple alleles	probabilities
heredity	mutations	Punnett squares
modern genetics	polygenic traits	

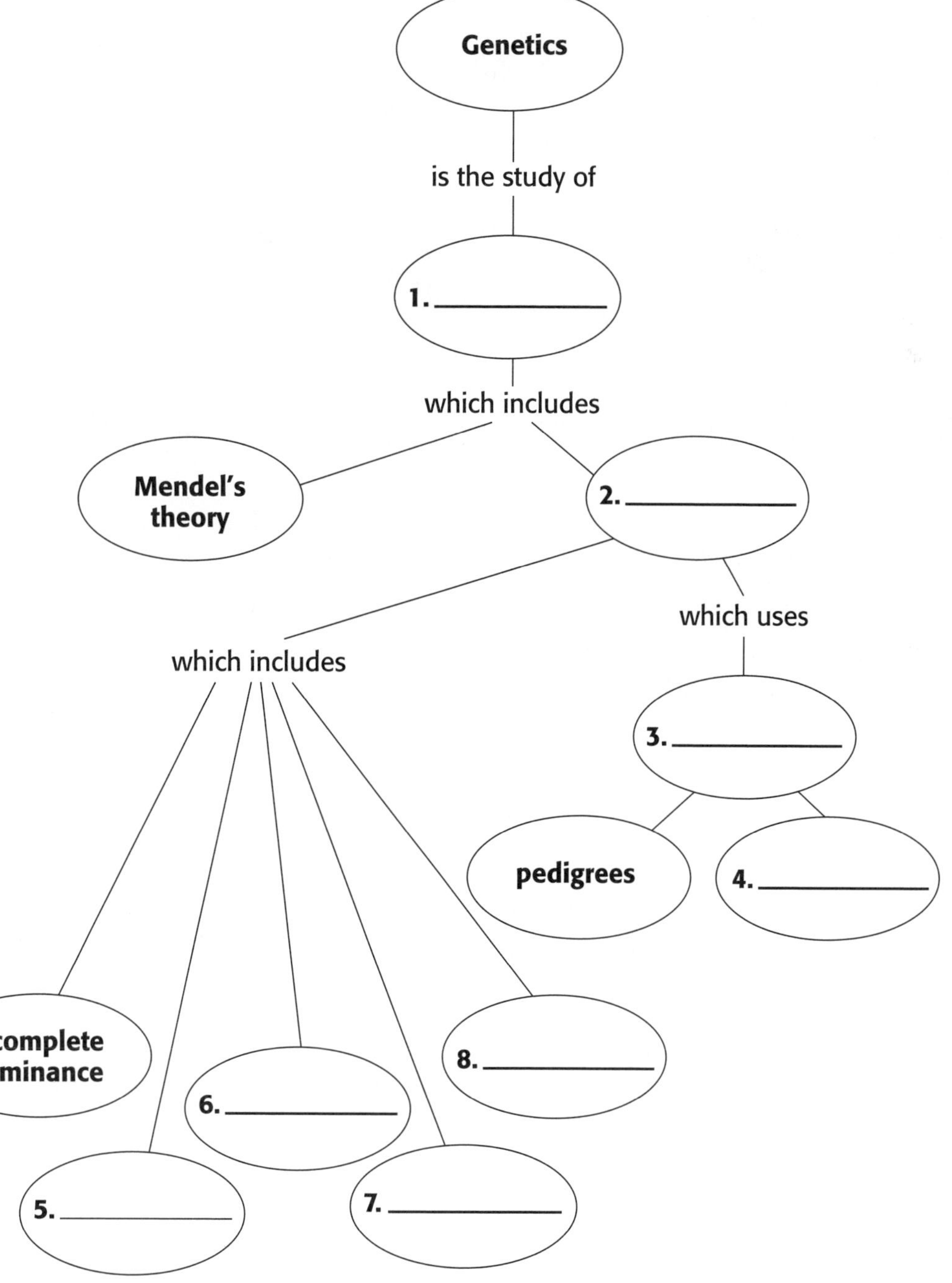

Critical Thinking

Work-Alikes

In the space provided, write the letter of the term or phrase that best describes how each numbered item functions.

_______ **1.** true-breeding

_______ **2.** self-fertilization

_______ **3.** cross-pollination

_______ **4.** probability

_______ **5.** alleles

_______ **6.** ratio

a. a musician who continues to play the same song

b. different versions of the same story

c. feeding food to someone else

d. predicting the outcome of a coin toss

e. feeding food to yourself

f. a baseball player's batting average

Cause and Effect

In the space provided, write the letter of the term or phrase that best matches each cause or effect given below.

Cause	Effect
7. _______________	heterozygous condition
8. _______________	small, matures quickly, produces many offspring
9. P generation cross	_______________
10. _______________	genetic disorder
11. test cross	_______________
12. incomplete dominance	_______________
13. _______________	sickle cell anemia
14. _______________	white fur in winter

a. a defective hemoglobin

b. garden pea plant for heredity studies

c. mutated genes

d. alleles of a particular gene are different

e. wavy-haired child

f. genotype of dominant individual determined

g. F_1 generation

h. pigment gene does not work in cold temperatures

| Critical Thinking *continued*

Linkages

In the spaces provided, write the letters of the two terms or phrases that are linked together by the term or phrase in the middle. The choices can be placed in any order. Some choices may be used more than once.

15. _______ use of mathematics _______

16. _______ F_1 _______

17. _______ law of independent _______ assortment

18. _______ multiply two separate _______ probabilities

19. _______ X chromosome _______

20. _______ incomplete _______ dominance

a. heterozygous condition

b. recessive allele

c. separation during meiosis

d. P

e. Mendel's experimentation

f. two alleles for different traits

g. blending of traits

h. unknown probability of a genetic cross

i. sex-linked trait

j. science of genetics

k. known probability of genetic cross

l. F_2

Analogies

An analogy is a relationship between two pairs of terms or phrases written as a : b :: c : d. The symbol : is read as "is to," and the symbol :: is read as "as." In the space provided, write the letter of the pair of terms or phrases that best completes the analogy shown.

_______ **21.** *RR* : homozygous dominant ::
  a. *Rr* : heterozygous
  b. *rr* : heterozygous
  c. *Yy* : homozygous recessive
  d. *yy* : heterozygous

_______ **22.** phenotype : physical appearance ::
  a. genotype : value
  b. genotype : set of alleles
  c. genotype : cover
  d. genotype : sex linkage

_______ **23.** law of segregation : alleles for the same trait ::
  a. codominance : one trait expressed
  b. self-fertilization : cross-pollination
  c. allele pairs : law of segregation
  d. law of independent assortment : alleles for different traits

_______ **24.** sex-linked traits : on X chromosomes ::
  a. autosomes : on X chromosomes
  b. autosomes : on X-linked traits
  c. autosomal traits : on X chromosomes
  d. inherited disorders : in carriers

Test Prep Pretest

In the space provided, write the letter of the term or phrase that best completes each statement or best answers each question.

_______ **1.** *Pisum sativum*, the garden pea, is a good subject to use in studying heredity for all of the following reasons EXCEPT
  a. Several varieties of *Pisum sativum* are available that differ in easily distinguishable traits.
  b. *Pisum sativum* is a small, easy-to-grow plant.
  c. *Pisum sativum* matures quickly and produces a large number of offspring.
  d. A *Pisum sativum* plant with male reproductive parts must cross-pollinate with a plant having female reproductive parts for reproduction to take place.

_______ **2.** Step 1 of Mendel's garden pea experiment, allowing each variety of garden pea to self-pollinate for several generations, produced the
  a. F_1 generation.   **c.** P generation.
  b. F_2 generation.   **d.** P_2 generation.

_______ **3.** In the F_2 generation in Mendel's experiments, the ratio of dominant to recessive phenotypes was
  a. 1:3.   **c.** 2:1.
  b. 1:2.   **d.** 3:1.

_______ **4.** The trait that was expressed in the F_1 generation in Mendel's experiment is considered
  a. recessive.   **c.** second filial.
  b. dominant.   **d.** parental.

_______ **5.** Mendel's law of segregation states that
  a. pairs of alleles are dependent on one another when separation occurs during gamete formation.
  b. pairs of alleles separate independently of one another after gamete formation.
  c. each pair of alleles remains together when gametes are formed.
  d. the two alleles for a trait separate when gametes are formed.

_______ **6.** A series of genetic crosses results in 787 long-stemmed plants and 277 short-stemmed plants. The probability that you will obtain short-stemmed plants if you repeat this experiment is
  a. $\dfrac{277}{1,064}$.   **c.** $\dfrac{787}{277}$.
  b. $\dfrac{277}{787}$.   **d.** $\dfrac{787}{1,064}$.

_______ **7.** Crossing a snapdragon that has red flowers with one that has white flowers produces a snapdragon that has pink flowers. The trait for flower color exhibits
- **a.** multiple alleles.
- **b.** complete dominance.
- **c.** incomplete dominance.
- **d.** codominance.

_______ **8.** Which of the following is NOT considered a genetic disorder?
- **a.** sickle cell anemia
- **b.** hemophilia
- **c.** AIDS
- **d.** cystic fibrosis

_______ **9.** On which of the following chromosomes would a sex-linked trait most likely be found in humans?
- **a.** X
- **b.** Y
- **c.** O
- **d.** YO

_______ **10.** The roan color of a horse is an example of
- **a.** homozygous alleles.
- **b.** codominance.
- **c.** incomplete dominance.
- **d.** Both (a) and (c)

Questions 11 and 12 refer to the figure at right, which represents a monohybrid cross between two individuals that are heterozygous for a trait.

	D	*d*
D	*DD*	*Dd*
__	*D__*	*d__*

_______ **11.** If the resulting phenotypic ratio is 3:1, the missing parental allele is
- **a.** *d*.
- **b.** *D*.
- **c.** *Dd*.
- **d.** *DD*.

_______ **12.** The two unknown genotypes in the offspring are
- **a.** *DD* and *dd*.
- **b.** *Dd* and *Dd*.
- **c.** *dd* and *DD*.
- **d.** *Dd* and *dd*.

_______ **13.** Which of the following summarizes one of Mendel's major hypotheses developed from his studies of garden peas?
- **a.** All of an individual's alleles make up its genotype.
- **b.** Traits that are intermediate between two parents are caused by genes that are incompletely dominant.
- **c.** There are alternative versions of genes, which are called alleles today.
- **d.** When two dominant alleles are expressed together, they are called codominant.

_______ **14.** Which of the following is an example of a test cross?
- **a.** *YY* × *YY*
- **b.** *YY* × *yy*
- **c.** *Yy* × *Yy*
- **d.** All of the above

| Test Prep Pretest *continued*

______15. What is the goal of genetic counseling?
    a. to cure genetic disorders
    b. to inform people about genetic disorders that could affect them or their offspring
    c. to use gene technology to correct certain genetic disorders
    d. to identify people who have a family history of genetic disorders

Question 16 refers to the figure below, which shows the inheritance of sickle cell anemia in a family.

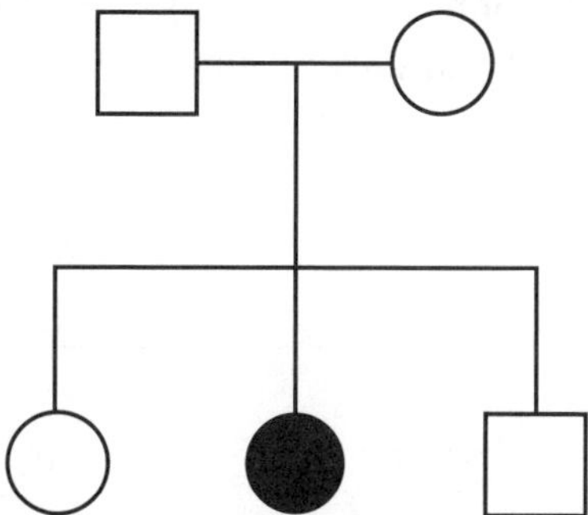

______16. Which of the following is true based on the information provided in the pedigree?
    a. Both parents have sickle cell anemia.
    b. Both parents carry an allele for sickle cell anemia.
    c. Sickle cell anemia is caused by a dominant allele.
    d. All three children are carriers of a defective gene that causes sickle cell anemia.

Complete each statement by writing the correct term or phrase in the space provided.

17. The investigator whose studies formed the basis of modern genetics is

_______________________ .

18. The _______________________ , or physical appearance, of an individual is

determined by the alleles that code for traits. The set of alleles that an individ-

ual has is called its _______________________ .

19. A cross between a pea plant that is true-breeding for green pod color
and one that is true breeding for yellow pod color is an example of

a(n) _______________________ cross.

Name _______________________________ Class _______________ Date _______________

20. Characteristics such as eye color, height, weight, and hair and skin color are

examples of _____________________ _____________________ because
several genes act together to influence a trait.

21. Mutations in genetic material may cause _____________________

_____________________, such as cystic fibrosis and muscular dystrophy.

Read each question, and write your answer in the space provided.

22. What approximate ratio of plants expressing contrasting traits did Mendel
calculate in his F_2 generation of garden peas? What steps did he take to
calculate this ratio?

____

____

____

____

23. Name Mendel's two laws of heredity.

____

____

24. Give an example of how the environment might influence gene expression.

____

____

____

25. Describe one sex-linked genetic disorder.

____

____

____

Quiz

Section: The Origins of Genetics

In the space provided, write the letter of the term or phrase that best completes each statement or best answers each question.

_______ **1.** The garden pea is a good subject for genetic study because it
 a. is easy to grow.
 b. has traits in two clearly different forms.
 c. produces many offspring.
 d. All of the above

_______ **2.** Mendel first allowed each variety of garden pea to self-pollinate for several generations. This produced the
 a. F_1 generation.
 c. P generation.
 b. F_2 generation.
 d. filial generation.

_______ **3.** What did Mendel find in the F_1 generation?
 a. a ratio of 3:1
 b. a ratio of 3:1:3
 c. 30 percent of the plants expressed the dominant trait
 d. 100 percent of the plants expressed the dominant trait

_______ **4.** Which ratio of plants expressing contrasting traits did Mendel find in the F_2 generation?
 a. 3:1.
 c. 3:3
 b. 3:1:3.
 d. 1:1

_______ **5.** The trait that disappeared in the F_1 generation
 a. reappeared in some plants in the F_2 generation.
 b. reappeared in all plants in the F_2 generation.
 c. reappeared in some plants in the P generation.
 d. was lost forever.

In the space provided, write the letter of the description that best matches the term or phrase.

_______ **6.** Mendel

_______ **7.** ratio

_______ **8.** T. A. Knight

_______ **9.** heredity

_______ **10.** genetics

a. comparison of two numbers

b. branch of biology that focuses on heredity

c. British farmer who crossed purple flowering pea plants with white flowering pea plants

d. passing traits from parents to offspring

e. developed rules that accurately predicted patterns of heredity

Assessment

Quiz

Section: Mendel's Theory

In the space provided, write the letter of the term or phrase that best completes each statement or best answers each question.

_______ **1.** Which of the following is a genotype for an individual heterozygous for a trait?
  a. *Rr*
  b. *tt*
  c. *YY*
  d. *Pr*

_______ **2.** Which of the following is NOT one of Mendel's major hypotheses?
  a. An individual receives two copies of a gene for each trait.
  b. Genes have alternative versions, which we now call alleles.
  c. Gametes carry several alleles for each inherited trait.
  d. When two alleles appear together, one may be dominant.

_______ **3.** Mendel's law of segregation states that the two alleles for a trait
  a. are linked to sex chromosomes during gamete formation.
  b. separate independently of one another after gamete formation.
  c. remain together when gametes are formed.
  d. separate when gametes are formed.

_______ **4.** Which of the following is a phenotype for pea plant heterozygous flower color?
  a. white flowers
  b. purple flowers
  c. pink flowers
  d. speckled flowers

In the space provided, write the letter of the description that best matches the term or phrase.

_______ **5.** homozygous

_______ **6.** heterozygous

_______ **7.** alleles

_______ **8.** genotype

_______ **9.** dominant

_______ **10.** recessive

a. form of a trait that is not expressed

b. different versions of a gene

c. for example, *pp*

d. set of all the alleles an individual has

e. for example, *Pp*

f. expressed form of a trait

Quiz

Section: Studying Heredity

In the space provided, write the letter of the term or phrase that best completes each statement or best answers each question.

______ **1.** In a pedigree, two normal parents produce a child with a genetic disorder. Based on this information, you know that the genetic disorder is caused by a
   a. sex-linked allele.
   b. dominant allele.
   c. codominant allele.
   d. recessive allele.

______ **2.** The results of a test cross are some purple-flowering plants and some white-flowering plants. What is the genotype of the original purple-flowering pea plant?
   a. *PP*
   b. *pp*
   c. *Pp*
   d. There is not enough information.

______ **3.** In a pedigree, a carrier would be represented as a(n)
   a. normal individual.
   b. individual who has a genetic disorder.
   c. square.
   d. circle.

______ **4.** Which of the following is NOT true of most sex-linked traits?
   a. located on the X chromosome   **c.** usually recessive
   b. located on autosomes   **d.** usually seen in males

In the space provided, write the letter of the description that best matches the term or phrase.

______ **5.** homozygous recessive

______ **6.** 1 *WW* : 2 *Ww* : 1 *ww*

______ **7.** Punnett square

______ **8.** test cross

______ **9.** probability

______ **10.** all *Ww* offspring

a. result of a cross between the parents *WW* and *ww*

b. result from a cross between the parents *Ww* and *Ww*

c. likelihood that an event will occur

d. used to determine the genotype of a purple-flowering pea plant

e. diagram used to predict the outcome of a genetic cross

f. good choice of genotype for a test cross to find out the genotype of an individual who might be *Ww* or *WW*

Assessment

Quiz

Section: Complex Patterns of Heredity

In the space provided, write the letter of the term or phrase that best completes each statement or best answers each question.

_______ **1.** When several genes influence a trait, the trait is said to be
    a. polygenic.   **c.** codominant.
    b. incompletely dominant.   **d.** completely dominant.

_______ **2.** Which of the following patterns of heredity can result in an intermediate trait, such as pink snapdragon flowers?
    a. multiple alleles
    b. incomplete dominance
    c. codominance
    d. sex-linked alleles

_______ **3.** Which of the following is responsible for the color of hydrangea flowers, which depends on the pH of the soil in which they are grown?
    a. multiple alleles
    b. incomplete dominance
    c. codominance
    d. environmental conditions

_______ **4.** Which of the following genetic disorders is caused by a sex-linked allele?
    a. sickle cell anemia   **c.** hemophilia A
    b. hypercholesterolemia   **d.** Tay-Sachs disease

In the space provided, write the letter of the description that best matches the term or phrase.

_______ **5.** sickle cell anemia

_______ **6.** Huntington's disease

_______ **7.** gene therapy

_______ **8.** genetic counseling

_______ **9.** mutation

_______ **10.** multiple alleles

a. determine the different ABO blood types

b. caused by a mutated allele that results in a defective hemoglobin protein

c. caused by a dominant allele located on an autosome

d. changes in DNA that can cause genetic disorders

e. informing people about genetic problems they or their offspring might have

f. replacing defective genes with copies of healthy genes, using gene technology

Chapter Test

Mendel and Heredity

In the space provided, write the letter of the term or phrase that best completes each statement or best answers each question.

______ **1.** When two different alleles occur together, the one that is expressed is called
  a. dominant.
  b. phenotypic.
  c. recessive.
  d. superior.

______ **2.** An organism that has inherited two of the same alleles of a gene from its parents is ______ for that trait.
  a. hereditary
  b. heterozygous
  c. homozygous
  d. a mutation

______ **3.** The law of segregation states that
  a. alleles of a gene separate from each other during gamete formation.
  b. different alleles of a gene can never be found in the same organism.
  c. each gene of an organism ends up in a different gamete.
  d. each gene is found on a different molecule of DNA.

______ **4.** The law of independent assortment applies only to genes that are
  a. sex-linked.
  b. located on different chromosomes or are far apart on the same chromosome.
  c. located on the same chromosome.
  d. autosomal.

______ **5.** Both sickle cell anemia and hemophilia
  a. are caused by genes coding for defective proteins.
  b. are seen in homozygous dominant individuals.
  c. provide resistance to malaria infections.
  d. are sex-linked traits.

______ **6.** If a characteristic is sex-linked, it
  a. occurs most commonly in males.
  b. occurs only in females.
  c. can never occur in females.
  d. is always fatal.

______ **7.** Which of the following is an example of gene therapy?
  a. A genetic counselor studies a pedigree.
  b. A student studies the colors of flowers in pea plants.
  c. A geneticist explains the inheritance of albinism by using a Punnett square.
  d. A physician transfers a normal gene into the DNA of a person with a mutated form of the gene.

Chapter Test *continued*

Questions 8–11 refer to the figure below, which shows a cross between two rabbits. In rabbits, black fur (B) is dominant to brown fur (b).

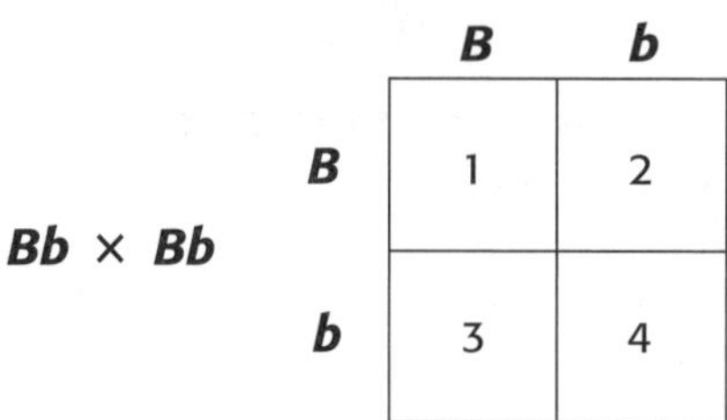

_______ **8.** The device illustrated above, which is used to organize genetic analysis, is called a
    a. Mendelian box.   **c.** genetic graph.
    b. Punnett square.   **d.** phenotypic paradox.

_______ **9.** The fur in both of the parents in the cross is
    a. black.   **c.** homozygous dominant.
    b. brown.   **d.** homozygous recessive.

_______ **10.** The phenotype of the offspring indicated by Box 3 would be
    a. brown.
    b. a mixture of brown and black.
    c. black.
    d. The phenotype cannot be determined.

_______ **11.** The genotype ratio ($BB : Bb$) of the F_1 generation would be
    a. 1:1.   **c.** 1:3.
    b. 3:1.   **d.** None of the above

Complete each statement by writing the correct term or phrase in the space provided.

_______ **12.** Which of the following investigators carried out studies that formed the basis of modern genetics?
    a. T. A. Knight
    b. Gregor Mendel
    c. Thomas Morgan
    d. Charles Darwin

_______ **13.** The garden pea is a good subject for genetic study because it
    a. clearly shows codominance.
    b. has traits in two clearly different forms.
    c. produces few offspring.
    d. has separate male and female flowers.

Chapter Test *continued*

_______**14.** The first step in Mendel's garden pea experiments was to
   a. remove the stamens of flowers on the plants.
   b. cross-pollinate two P generation plants with contrasting traits.
   c. allow each plant variety to self-pollinate for several generations.
   d. allow the F_1 generation to self-pollinate.

_______**15.** Which ratio of dominant to recessive phenotypes did Mendel find in his F_2 generation?
   a. 1:3   **c.** 2:1
   b. 3:1   **d.** 4:1

_______**16.** Black fur is dominant to brown fur in rabbits. White and gray fur exhibit incomplete dominance. How can you find out the genotype of a rabbit with black fur?
   a. Mate the black rabbit with a white rabbit.
   b. Mate the black rabbit with a another black rabbit.
   c. Mate the black rabbit with a gray rabbit.
   d. Mate the black rabbit with a brown rabbit.

_______**17.** Which of the following is true when analyzing a pedigree?
   a. If a disorder is caused by a recessive trait, every offspring afflicted with the disorder will have a parent with the disorder.
   b. If a disorder is caused by a dominant trait, two normal parents can produce an offspring with the disorder.
   c. If a disorder is caused by a recessive trait, the normal parents of every offspring with the disorder are carriers.
   d. If a disorder is caused by a sex-linked trait, only male offspring will have the disorder.

_______**18.** Genetic disorders are caused by
   a. faulty proteins.   **c.** damaged genes.
   b. genetic mutations.   **d.** All of the above

_______**19.** Which of the following human genetic disorders is caused by a dominant allele?
   a. cystic fibrosis   **c.** Huntington's disease
   b. hemophilia   **d.** sickle cell anemia

_______**20.** What do genetic counselors do?
   a. They look for cures for deadly genetic disorders.
   b. They try to replace defective genes with healthy ones, using an approach called gene therapy.
   c. They inform people about genetic disorders that could affect them.
   d. All of the above

Chapter Test

Mendel and Heredity

In the space provided, write the letter of the description that best matches the term or phrase.

______ **1.** Gregor Mendel

______ **2.** 3:1

______ **3.** homozygous dominant

______ **4.** homozygous recessive

______ **5.** T. A. Knight

______ **6.** genotype

______ **7.** law of segregation

______ **8.** law of independent assortment

______ **9.** genetic disorder

______ **10.** $1\,PP : 2\,Pp : 1\,pp$

a. genotype of an individual used in a test cross

b. harmful effects produced by inherited genetic mutations

c. performed breeding experiments on pea plants 200 years before Mendel did

d. two alleles for a trait separate when gametes are formed

e. person whose studies formed the basis of modern genetics

f. alleles for different genes separate independently of one another during gamete formation

g. an individual's set of alleles

h. the genotype PP is an example

i. genotypic ratio expected when crossing two individuals heterozygous for a trait

j. phenotypic ratio expected when crossing two individuals heterozygous for a trait

Complete each statement by writing the correct term or phrase in the space provided.

11. When two members of the F_1 generation are crossed, the offspring are

referred to as the _____________________ generation.

12. A trait that is determined by a gene found only on the X chromosome is said

to be a(n) _____________________ trait.

13. Parents with genotypes Pp and pp can produce offspring of genotypes

_____________________ and _____________________ .

14. Crossing two pea plants heterozygous for flower color should produce a

phenotypic ratio of _____________________ in the offspring.

Name _________________________________ Class _______________ Date _______________

15. In a pedigree, if two normal parents produce a child with a genetic disorder, then the disorder is caused by a(n) _____________________ allele.

16. Providing medical guidance about genetic disorders that could affect patients or their offspring is called _____________________ _____________________ .

17. Mendel removed the stamens on pea flowers in order to _____________________-_____________________ two P generation plants that had contrasting forms of a trait.

18. Mendel allowed each variety of garden pea to _____________________-_____________________ for several generations. This ensured that each plant in the P generation was _____________________-_____________________ for a particular trait.

Read each question, and write your answer in the space provided.

19. Explain how polygenic traits occur.

____

____

____

20. Explain why sex-linked traits are more common in males than in females.

____

____

____

____

21. Give two examples of traits that are influenced by the environment.

____

____

____

____

Chapter Test *continued*

22. Explain the causes and symptoms of cystic fibrosis.

____

____

____

____

23. List the four major hypotheses Mendel developed using his garden pea experiments.

____

____

____

24. List four characteristics that make the garden pea a good subject for genetic study.

____

____

____

25. In step 1 of his experiments, how did Mendel ensure that each variety of garden pea was true-breeding for a particular trait?

____

____

____

____

(Math Lab)   **DATASHEET FOR IN-TEXT LAB**

Calculating Mendel's Ratios

Background

You can calculate the ratios Mendel obtained in the F_2 generation for the traits he studied.

Data Table

Contrasting traits	F_2 generation results		Ratio
Flower color	705 purple	224 white	3.15:1
Seed color	6,022 yellow	2,001 green	
Seed shape	5,474 round	1,850 wrinkled	
Pod color	428 green	152 yellow	
Pod shape	882 round	299 constricted	
Flower position	651 axial	207 top	
Plant height	787 tall	277 dwarf	

Analysis

1. Calculate the ratio for each contrasting trait to complete the table above. Use colon form.

____

____

2. State the ratio for each contrasting trait in words and as a fraction.

____

____

____

____

____

3. Critical Thinking
Interpreting Results Do the data confirm a 3:1 ratio in the F_2 generation for each of the traits he studied?

____

Identifying Dominant or Recessive Traits

You can determine some of the genotypes and all of the phenotypes for human traits that are inherited as simple dominant or recessive traits.

MATERIALS

- pencil
- paper

Data Table	
Dominant trait	**Recessive trait**
Cleft chin	No cleft
Dimples	No dimples
Hair above knuckles	Hairless fingers
Freckles	No freckles

Procedure

1. Look at the table above. For each trait, circle the phenotype that best matches your own phenotype.
2. Determine how many students in your class share your phenotype by recording your results in a table on the chalkboard.

Analysis

1. **Summarize** the class results for each trait.

____

____

2. **Calculate** the class dominant:recessive ratio for each trait.

____

____

3. **Critical Thinking**
   Applying Information For which phenotypes in the table can you determine a person's genotype without ever having seen his or her parents? Explain.

____

____

Data Lab

Analyzing a Test Cross

DATASHEET FOR IN-TEXT LAB

Background

You can use a test cross to determine whether a plant with purple flowers is heterozygous (*Pp*) or homozygous dominant (*PP*). Fill in the boxes in each Punnett square shown below.

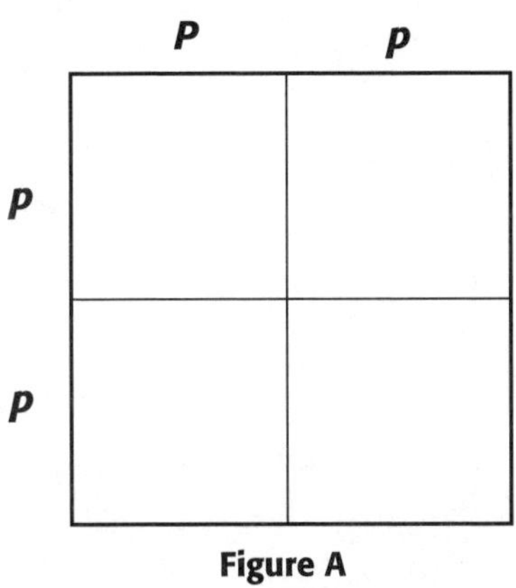

Figure A
Heterozygous (*Pp*) plant

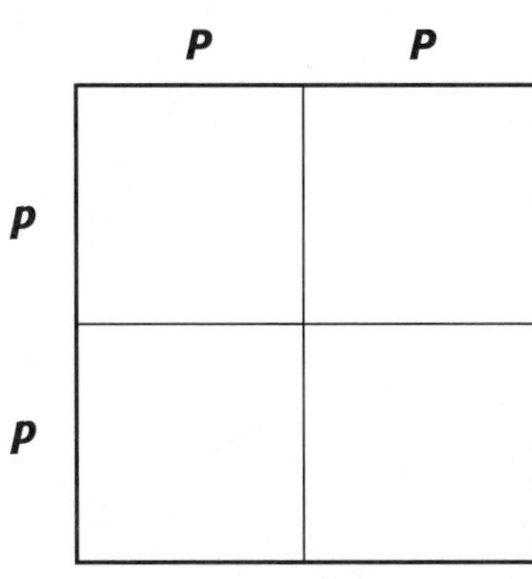

Figure B
Homozygous (*PP*) plant

Analysis

1. Determine what the letters at the top and side of each box represent.

____

2. Determine what the letters in each box represent.

____

3. Calculate the genotypic and phenotypic ratios that would be predicted if the parent of the unknown genotype were homozygous for the trait (Figure B).

____

____

4. Critical Thinking
   Predicting Outcomes If half of the offspring have white flowers, what is the genotype of the plant with purple flowers?

____

Predicting the Results of Crosses Using Probabilities

Background

In rabbits, the allele B for black hair is dominant over the allele b for brown hair. You can practice using probabilities to predict the outcome of genetic crosses by completing the genetic problems below. Draw Punnett squares for each problem.

Analysis

1. **Calculate** the probability of homozygous dominant (*BB*) offspring resulting from a cross between two heterozygous (*Bb*) parents.

2. **Calculate** the probability of heterozygous offspring resulting from a cross between a heterozygous parent and a homozygous recessive (*bb*) parent.

3. **Calculate** the probability of heterozygous offspring resulting from a cross between a homozygous dominant parent and a homozygous recessive parent.

4. **Calculate** the probability of homozygous dominant offspring resulting from a cross between a heterozygous parent and a homozygous recessive parent.

Data Lab

Evaluating a Pedigree

Background

Pedigrees, such as the one below, can be used to track different genetic traits. Use the pedigree below to practice interpreting a pedigree.

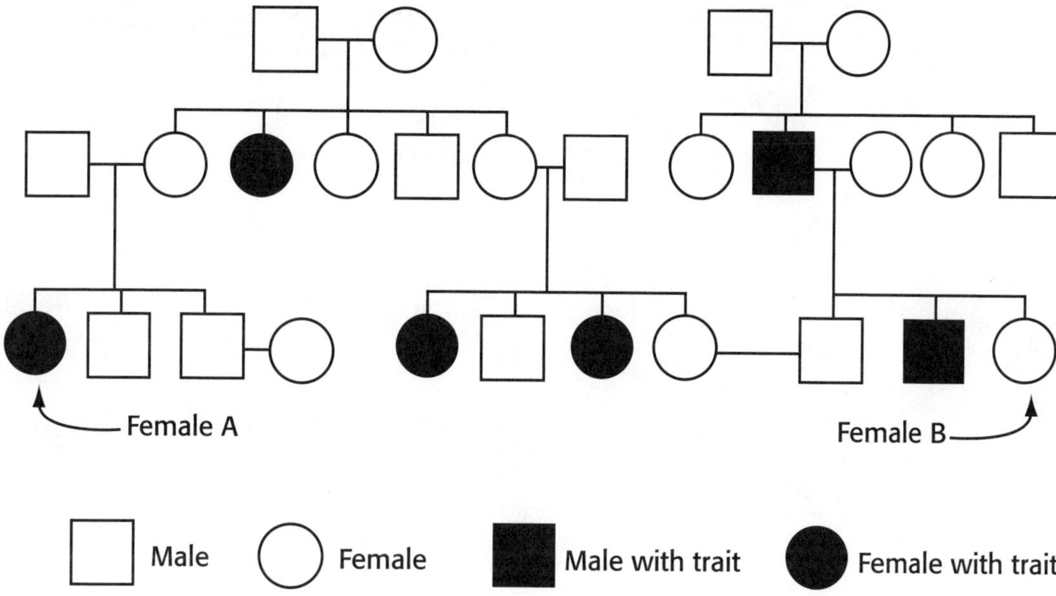

Analysis

1. **Interpret** the pedigree to determine whether the trait is sex-linked or autosomal and whether the trait is inherited in a dominant or recessive manner.

   ____

2. **Determine** whether Female A is homozygous or heterozygous.

   ____

3. **Critical Thinking**
   Applying Information If Female B has children with a homozygous individual, what is the probability that the children will be heterozygous?

   ____

DATASHEET FOR IN-TEXT LAB

Modeling Monohybrid Crosses

MATERIALS

- lentils
- green peas
- 2 Petri dishes

SKILLS

- Predicting outcomes
- Calculating data
- Organizing data
- Analyzing data

OBJECTIVES

- **Predict** the genotypic and phenotypic ratios of offspring resulting from the random pairing of gametes.
- **Calculate** the genotypic ratio and phenotypic ratio among the offspring of a monohybrid cross.

Before You Begin

A **monohybrid cross** is a cross that involves one pair of contrasting traits. Different versions of a gene are called alleles. When two different alleles are present and one is expressed completely and the other is not, the expressed allele is dominant and the unexpressed allele is recessive.

1. Write a definition for each boldface term in the paragraph above. Use a separate sheet of paper.

2. Based on the objectives for this lab, write a question you would like to explore about heredity.

____

____

Procedure
PART A: SIMULATING A MONOHYBRID CROSS

1. You will model the random pairing of alleles by choosing lentils and peas from Petri dishes. These dried seeds will represent the alleles for seed color. A green pea will represent G, the dominant allele for green seeds, and a lentil will represent g, the recessive allele for yellow seeds.

 55

2. Each Petri dish will represent a parent. Label one Petri dish "female gametes" and the other Petri dish "male gametes." Place one green pea and one lentil in the Petri dish labeled "female gametes" and place one green pea and one lentil in the Petri dish labeled "male gametes."

3. Each parent contributes one allele to each offspring. Model a cross between these two parents by choosing a random pairing of the dried seeds from the two Petri dishes. Do this by simultaneously picking one seed from each Petri dish without looking. Place the pair of seeds together on the lab table. The pair of seeds represents the genotype of one offspring.

4. Record the genotype of the first offspring in Table A below.

Table A		
Gamete Pairings		
Trial	**Offspring genotype**	**Offspring phenotype**
1		
2		
3		
4		
5		
6		
7		
8		
9		
10		

5. Return the seeds to their original dishes and repeat step 3 nine more times. Record the genotype of each offspring in Table A.

6. Based on each offspring's genotype, determine and record each offspring's phenotype.

____

____

____

PART B: CALCULATING GENOTYPIC AND PHENOTYPIC RATIOS

7. You will be using Table B to record your data.

8. Determine the genotypic and phenotypic ratios among the offspring. First count and record in Table B the number of homozygous dominant, heterozygous, and homozygous recessive individuals you recorded in Table A. Then

Modeling Monohybrid Crosses *continued*

record the number of offspring that produce green seeds and the number that produce yellow seeds under "Phenotypes" in Table B.

Table B		
Offspring Ratios		
Genotypes	**Total**	**Genotypic ratios**
Homozygous dominant *(GG)*		
Heterozygous *(Gg)*		______ : ______ : ______
Homozygous recessive *(gg)*		
Phenotypes		**Phenotypic ratios**
Green seeds		
Yellow seeds		

9. Calculate the genotypic ratio for each genotype using the following equation:

$$\text{phenotypic ratio} = \frac{\text{number of offspring with a given genotype}}{\text{total number of offspring}}$$

10. Calculate the phenotypic ratio for each phenotype using the following equation:

$$\text{phenotypic ratio} = \frac{\text{number of offspring with a given phenotype}}{\text{total number of offspring}}$$

11. Now pool the data for the whole class, and record the data in Table C below.

12. Compare the class's sample with your small sample of 10. Calculate the genotypic and phenotypic ratios for the class data, and record them in Table C below.

13. Construct a Punnett square showing the parents and their offspring in your lab report.

14. Clean up your materials before leaving the lab.

Table C		
Offspring Ratios (Class Data)		
Genotypes	**Total**	**Genotypic ratios**
Homozygous dominant *(GG)*		
Heterozygous *(Gg)*		______ : ______ : ______
Homozygous recessive *(gg)*		
Phenotypes		**Phenotypic ratios**
Green seeds		
Yellow seeds		

Modeling Monohybrid Crosses *continued*

Analyze and Conclude

1. Summarizing Results What trait is being studied in this investigation?

____

2. Analyzing Data What are the genotypes of the parents? Describe the genotypes of both parents using the terms *homozygous* or *heterozygous*, or both. Did Table B reflect a classic monohybrid-cross phenotypic ratio of 3:1?

____

____

____

____

3. Drawing Conclusions If a genotypic ratio of 1:2:1 is observed, what must the genotypes of both parents be?

____

4. Predicting Patterns Show what the genotypes of the parents would be if 50 percent of the offspring were green and 50 percent of the offspring were yellow.

____

____

5. Further Inquiry Construct a Punnett square for the cross of a heterozygous black guinea pig and an unknown guinea pig whose offspring include a recessive white-furred individual. Use a separate sheet of paper. What are the possible genotypes of the unknown parent?

Interpreting Information in a Pedigree

Organizing information is often the key to solving a problem. Tracing the hereditary characteristics over many generations can be confusing unless the information is well organized. In this lab, you will learn how to organize hereditary information, making it much easier to analyze.

OBJECTIVES

Analyze a pedigree.

Construct a pedigree.

MATERIALS

- paper
- pencil

Procedure

1. Examine Pedigree I, which traces the dimples trait through three generations of a family. Blackened symbols represent people with dimples. Circles represent females, and squares represent males.

FIGURE 1

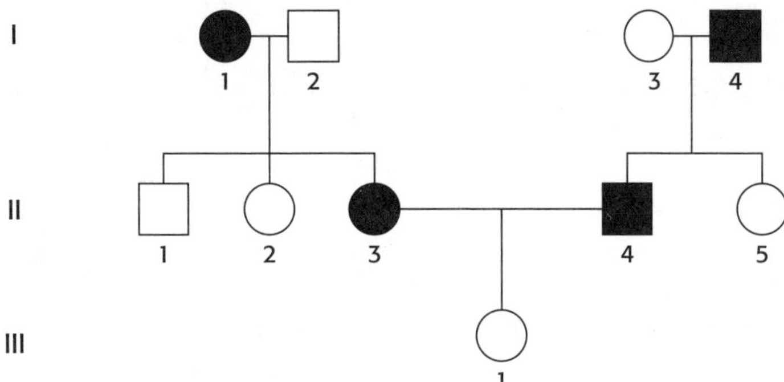

Pedigree I

2. Read the following passage, which describes the family shown in Pedigree I. Write the name of each person below the correct symbol in Pedigree I.

   Although Jane and Joe Smith have dimples, their daughter, Clarissa, does not. Joe's father has dimples, but his mother, and his sister, Grace, do not. Jane's father, Mr. Renaldo, her brother, Jorge, and her sister, Emily, do not have dimples, but her mother does.

3. Look at Pedigree I again.

   - How are marriage and offspring symbolized?

   ____

   ____

Interpreting Information in a Pedigree *continued*

- What do the Roman numerals symbolize?

  ____

4. Construct a pedigree based on the following passage about curly hair.

Andy, Penny, and Delbert have curly hair, but their mother, Mrs. Cummins, does not. Mrs. Giordano, Mrs. Cummin's sister, has curly hair, but her parents, Mr. & Mrs. Lutz, do not. Deidra and Darlene Giordano have curly hair, but their sister, Katie, like her father, has straight hair.

Analysis and Conclusions

1. Summarizing Observations What type of information does a pedigree contain?

____

____

2. Evaluating Models What advantages does a pedigree have over a written passage?

____

____

3. Interpreting Information Take another look at Pedigree I. A genetic counselor analyzing Pedigree I suggests that a person only needs to have one dominant allele for dimples (D) in order to have dimples. If this is true, what is the genotype of person 1 in the third generation of Pedigree I?

____

____

____

Analyzing Corn Genetics

In this lab, you will design and conduct an experiment to compare the germination rate and survival rate of corn from normal seeds with seeds containing the mutation for albinism.

OBJECTIVES

Develop a hypothesis to explain the poor yield of a corn crop.

Design and **conduct** an experiment to test your hypothesis.

Compare germination and survival rates of three lots of corn seeds.

Evaluate your results.

POSSIBLE MATERIALS

- corn seeds (10 each from lots A, B, and C)
- lab apron
- plant tray
- safety goggles
- soil, potting (3 kg)
- water

Finding Out More Information

In simple Mendelian genetics, a physical characteristic—called a *phenotype*—is controlled by two alleles. These two alleles reside on the chromosomes and are passed to the offspring through sexual reproduction, one allele being inherited from the father and one from the mother. The two alleles are considered the organism's *genotype*, and each allele may be dominant or recessive. A *dominant* phenotype will appear if an organism has a dominant genotype—both dominant alleles, or one dominant and one recessive allele. A *recessive* phenotype will appear only if an organism has both recessive alleles.

Albinism is a characteristic of some organisms in which the cells lack pigments. Albinism is a recessive trait, so an individual with two recessive alleles will show the trait. In mammals, physical characteristics of albinism may include white hair, pale skin, and lack of coloring in the irises of the eyes. The eyes may appear pink because the blood vessels at the back of the eye become visible.

In plants, albinism is characterized by the failure to produce *chlorophyll*, a plant pigment necessary for photosynthesis. Because the trait is recessive, parents with the normal phenotype may produce seeds with the alleles for albinism. However, the seeds do not develop because they lack chlorophyll. Thus, the presence of albinism in a corn crop may prevent a farmer from being able to maximize corn production.

- What are possible genotypes of normal corn plants?

____

____

• How can you determine the genotype of a normal corn plant?

____

____

____

____

Procedure

FORMING A HYPOTHESIS

Based on what you have learned, form a hypothesis about how albinism affects corn crop yield.

1. How can a farmer ensure maximum corn production?

____

____

____

2. Write your own hypothesis. A possible hypothesis might be "The presence of the recessive allele for albinism reduces corn yield by 25 percent."

____

____

COMING UP WITH A PLAN

Plan and conduct an experiment that will determine the germination and survival rates of three lots of corn seeds. Be sure to control variables among the three lots. Consult with your teacher about which variables you need to consider.

3. Write out a procedure for your experiment on a separate sheet of paper. As you plan the procedure, make the following decisions.

• Decide how far apart you will plant the seeds and how deep you will plant them.

• Decide how often and how much you will water the seeds.

• Decide where you will keep the seeds.

• Decide how long you will wait for the seeds to germinate and how long after germination you will measure growth.

• Decide what safety procedures are necessary. Add them to your written procedure.

4. Using graph paper or a computer, construct data tables to organize your data. You might choose tables similar to those shown below. Be sure that your tables fit your investigation.

5. Have your teacher approve your plan.

PERFORMING THE EXPERIMENT

6. Put on safety goggles and a lab apron.

7. Implement your plan, using the equipment and safety procedures that you selected.

8. Record your data in your tables. If necessary, revise your tables to include variables that you did not think of while planning your experiment.

9. At the end of the experiment, pool the class data and determine the germination rate and survival rate for each lot of corn seeds.

10. When you finish, clean and store your equipment. Recycle or dispose of all materials as instructed by your teacher. Wash your hands if you handled seeds or soil.

TABLE 1 GROUP DATA

Lot	Number of seeds	Number normal plants germinated	Number albino plants germinated	Number normal plants @ 4 wk	Number albino plants @ 4 wk
A	10				
B	10				
C	10				

TABLE 2 CLASS DATA

Lot	Number of seeds	Number normal plants germinated	Number albino plants germinated	Number normal plants @ 4 wk	Number albino plants @ 4 wk	Total germination rate (%)	Total germination rate (%)
A	50						
B	50						
C	50						

Analysis

1. **Analyzing Data** Using the class data, what was the germination rate for seeds in lot A? Lot B? Lot C?

2. **Analyzing Data** Using the class data, what was the survival rate for seeds in lot A? Lot B? Lot C? Explain any difference you observed.

3. **Analyzing Results** Did your results support your hypothesis? Explain.

4. **Recognizing Patterns** Use **Figure 1** to draw a Punnett square to show how albinism is inherited in corn plants. Use G to represent the allele for chlorophyll and g to represent the allele for albinism. Explain how you developed your Punnett square.

Analyzing Corn Genetics *continued*

FIGURE 1 PUNNETT SQUARE FOR ALBINISM IN CORN PLANTS

G		
g		

Conclusions

1. Evaluating Methods Did your experimental design give clear results? If not, how might you improve your design?

____

____

____

____

2. Interpreting Information What are the genotypes of the parents of an albino individual?

____

____

____

3. Drawing Conclusions What does the Punnett square reveal about crop yield?

____

____

____

4. Applying Conclusions How is the class data you and your classmates gathered in the experiment useful to a farmer planning a corn crop?

____

____

____

Name _______________________________ Class _______________ Date _____________

DATASHEET FOR IN-TEXT LAB

Calculating Mendel's Ratios

Background

You can calculate the ratios Mendel obtained in the F_2 generation for the traits he studied.

	Data Table		
Contrasting traits	**F_2 generation results**		**Ratio**
Flower color	705 purple	224 white	3.15:1
Seed color	6,022 yellow	2,001 green	
Seed shape	5,474 round	1,850 wrinkled	
Pod color	428 green	152 yellow	
Pod shape	882 round	299 constricted	
Flower position	651 axial	207 top	
Plant height	787 tall	277 dwarf	

Analysis

1. Calculate the ratio for each contrasting trait to complete the table above. Use colon form.

3:1, 2.96:1, 2.81:1, 2.95:1, 3.14:1, 2.84:1

2. State the ratio for each contrasting trait in words and as a fraction.

Three yellow seed to one green seed, $\frac{3}{1}$; three round to one wrinkled, $\frac{3}{1}$; three green pod to one yellow pod, $\frac{3}{1}$; three round pod to one constricted pod, $\frac{3}{1}$; three axial to one top, $\frac{3}{1}$; three tall to one dwarf, $\frac{3}{1}$

3. Critical Thinking
Interpreting Results Do the data confirm a 3:1 ratio in the F_2 generation for each of the traits he studied?

Yes.

Name _______________________________ Class _______________ Date _______________

DATASHEET FOR IN-TEXT LAB

Identifying Dominant or Recessive Traits

You can determine some of the genotypes and all of the phenotypes for human traits that are inherited as simple dominant or recessive traits.

MATERIALS

- pencil
- paper

Data Table	
Dominant trait	**Recessive trait**
Cleft chin	No cleft
Dimples	No dimples
Hair above knuckles	Hairless fingers
Freckles	No freckles

Procedure

1. Look at the table above. For each trait, circle the phenotype that best matches your own phenotype.

2. Determine how many students in your class share your phenotype by recording your results in a table on the chalkboard.

Analysis

1. **Summarize** the class results for each trait.

   Answers will vary.

   ____

2. **Calculate** the class dominant:recessive ratio for each trait.

   Answers will vary.

   ____

3. **Critical Thinking**
   Applying Information For which phenotypes in the table can you determine a person's genotype without ever having seen his or her parents? Explain.

   The recessive traits. Recessive traits must be homozygous to be expressed.

   ____

Name _______________________________ Class _______________ Date _____________

Data Lab **DATASHEET FOR IN-TEXT LAB**

Analyzing a Test Cross

Background

You can use a test cross to determine whether a plant with purple flowers is heterozygous (*Pp*) or homozygous dominant (*PP*). Fill in the boxes in each Punnett square shown below.

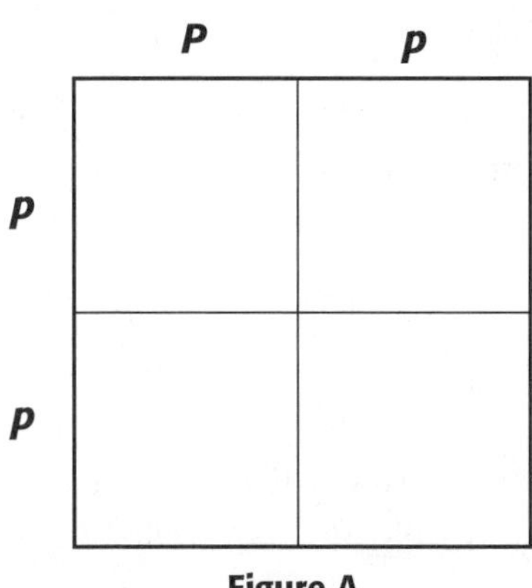

Figure A
Heterozygous (*Pp*) plant

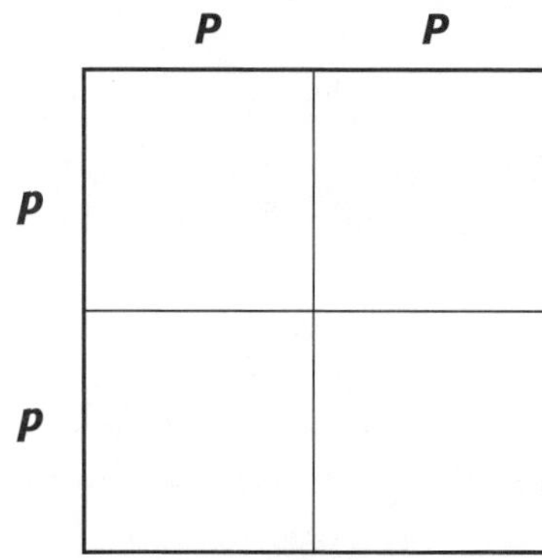

Figure B
Homozygous (*PP*) plant

Analysis

1. Determine what the letters at the top and side of each box represent.

possible alleles each parent can produce

2. Determine what the letters in each box represent.

the genotype of each possible kind of offspring

3. Calculate the genotypic and phenotypic ratios that would be predicted if the parent of the unknown genotype were homozygous for the trait (Figure B).

The genotypic ratio will be 4 *Pp* : 0 *PP* : 0 *pp*. The phenotypic ratio for the

offspring will be 4 purple : 0 white.

4. Critical Thinking
Predicting Outcomes If half of the offspring have white flowers, what is the genotype of the plant with purple flowers?

heterozygous

Name _____________________________ Class _____________ Date __________

  DATASHEET FOR IN-TEXT LAB

Predicting the Results of Crosses Using Probabilities

Background

In rabbits, the allele B for black hair is dominant over the allele b for brown hair. You can practice using probabilities to predict the outcome of genetic crosses by completing the genetic problems below. Draw Punnett squares for each problem.

Analysis

1. **Calculate** the probability of homozygous dominant (BB) offspring resulting from a cross between two heterozygous (Bb) parents.

   $\frac{1}{4}$

2. **Calculate** the probability of heterozygous offspring resulting from a cross between a heterozygous parent and a homozygous recessive (bb) parent.

   $\frac{1}{2}$

3. **Calculate** the probability of heterozygous offspring resulting from a cross between a homozygous dominant parent and a homozygous recessive parent.

   1

4. **Calculate** the probability of homozygous dominant offspring resulting from a cross between a heterozygous parent and a homozygous recessive parent.

   0

Name _________________________ Class _____________ Date __________

Data Lab

DATASHEET FOR IN-TEXT LAB

Evaluating a Pedigree

Background

Pedigrees, such as the one below, can be used to track different genetic traits.
Use the pedigree below to practice interpreting a pedigree.

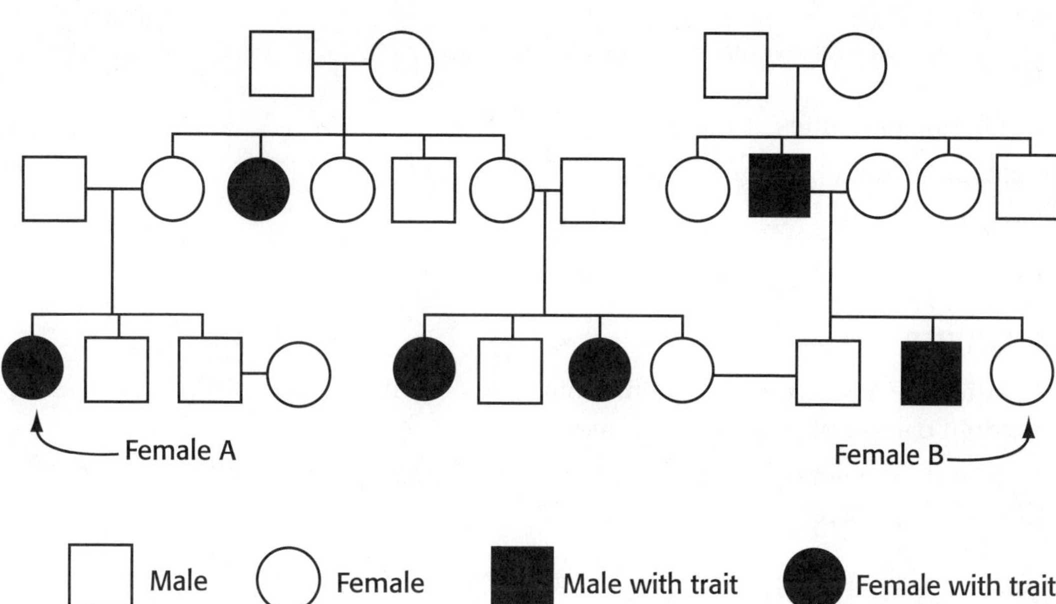

Analysis

1. **Interpret** the pedigree to determine whether the trait is sex-linked or autosomal and whether the trait is inherited in a dominant or recessive manner.

   autosomal recessive

2. **Determine** whether Female A is homozygous or heterozygous.

   homozygous

3. **Critical Thinking**
   Applying Information If Female B has children with a homozygous individual, what is the probability that the children will be heterozygous?

   $\frac{1}{2}$

Name _______________________________ Class _______________ Date _______________

DATASHEET FOR IN-TEXT LAB

Modeling Monohybrid Crosses

MATERIALS

- lentils
- green peas
- 2 Petri dishes

SKILLS

- Predicting outcomes
- Calculating data
- Organizing data
- Analyzing data

OBJECTIVES

- **Predict** the genotypic and phenotypic ratios of offspring resulting from the random pairing of gametes.
- **Calculate** the genotypic ratio and phenotypic ratio among the offspring of a monohybrid cross.

Before You Begin

A **monohybrid cross** is a cross that involves one pair of contrasting traits. Different versions of a gene are called alleles. When two different alleles are present and one is expressed completely and the other is not, the expressed allele is dominant and the unexpressed allele is recessive.

1. Write a definition for each boldface term in the paragraph above. Use a separate sheet of paper. **Answers appear in the TE for this lab.**

2. Based on the objectives for this lab, write a question you would like to explore about heredity.

   Answers may vary. ____

   ____

Procedure

PART A: SIMULATING A MONOHYBRID CROSS

1. You will model the random pairing of alleles by choosing lentils and peas from Petri dishes. These dried seeds will represent the alleles for seed color. A green pea will represent G, the dominant allele for green seeds, and a lentil will represent g, the recessive allele for yellow seeds.

Name _________________________________ Class ________________ Date ______________

Modeling Monohybrid Crosses *continued*

2. Each Petri dish will represent a parent. Label one Petri dish "female gametes" and the other Petri dish "male gametes." Place one green pea and one lentil in the Petri dish labeled "female gametes" and place one green pea and one lentil in the Petri dish labeled "male gametes."

3. Each parent contributes one allele to each offspring. Model a cross between these two parents by choosing a random pairing of the dried seeds from the two Petri dishes. Do this by simultaneously picking one seed from each Petri dish without looking. Place the pair of seeds together on the lab table. The pair of seeds represents the genotype of one offspring.

4. Record the genotype of the first offspring in Table A below.

Table A

Trial	Offspring genotype	Offspring phenotype
	Gamete Pairings	
1		
2		
3		
4		
5		
6		
7		
8		
9		
10		

5. Return the seeds to their original dishes and repeat step 3 nine more times. Record the genotype of each offspring in Table A.

6. Based on each offspring's genotype, determine and record each offspring's phenotype.

GG **5 green seeds;** *Gg* **5 green seeds;** *gg* **5 yellow seeds.**

____

____

PART B: CALCULATING GENOTYPIC AND PHENOTYPIC RATIOS

7. You will be using Table B to record your data.

8. Determine the genotypic and phenotypic ratios among the offspring. First count and record in Table B the number of homozygous dominant, heterozygous, and homozygous recessive individuals you recorded in Table A. Then

Name _________________________________ Class _______________ Date _____________

Modeling Monohybrid Crosses *continued*

record the number of offspring that produce green seeds and the number that produce yellow seeds under "Phenotypes" in Table B. **Answers appear in the TE for this lab.**

Table B		
Offspring Ratios		
Genotypes	**Total**	**Genotypic ratios**
Homozygous dominant *(GG)*		
Heterozygous *(Gg)*		_____ : _____ : _____
Homozygous recessive *(gg)*		
Phenotypes		**Phenotypic ratios**
Green seeds		
Yellow seeds		

9. Calculate the genotypic ratio for each genotype using the following equation:

$$\text{phenotypic ratio} = \frac{\text{number of offspring with a given genotype}}{\text{total number of offspring}}$$

10. Calculate the phenotypic ratio for each phenotype using the following equation:

$$\text{phenotypic ratio} = \frac{\text{number of offspring with a given phenotype}}{\text{total number of offspring}}$$

11. Now pool the data for the whole class, and record the data in Table C below.

12. Compare the class's sample with your small sample of 10. Calculate the genotypic and phenotypic ratios for the class data, and record them in Table C below.

13. Construct a Punnett square showing the parents and their offspring in your lab report.

14. Clean up your materials before leaving the lab.

Table C		
Offspring Ratios (Class Data)		
Genotypes	**Total**	**Genotypic ratios**
Homozygous dominant *(GG)*		
Heterozygous *(Gg)*		_____ : _____ : _____
Homozygous recessive *(gg)*		
Phenotypes		**Phenotypic ratios**
Green seeds		
Yellow seeds		

Name _______________________________ Class _______________ Date _______________

Modeling Monohybrid Crosses *continued*

Analyze and Conclude

1. Summarizing Results What trait is being studied in this investigation?

The trait being investigated is seed, or fruit, color.

2. Analyzing Data What are the genotypes of the parents? Describe the genotypes of both parents using the terms *homozygous* or *heterozygous*, or both. Did Table B reflect a classic monohybrid-cross phenotypic ratio of 3:1?

Both parents are heterozygous green, *Gg*. Each seed represents an allele.

The pairs represent the gametes that an offspring will receive. Answers will

vary. In a sample size of 10 crosses, no clear ratio may be evident. When

combining data from the entire class, the 3:1 ratio should be seen.

3. Drawing Conclusions If a genotypic ratio of 1:2:1 is observed, what must the genotypes of both parents be?

Both parents must be heterozygous, *Gg*.

4. Predicting Patterns Show what the genotypes of the parents would be if 50 percent of the offspring were green and 50 percent of the offspring were yellow.

One parent would be heterozygous, *Gg*, and the other parent would be

homozygous recessive, *gg*.

5. Further Inquiry Construct a Punnett square for the cross of a heterozygous black guinea pig and an unknown guinea pig whose offspring include a recessive white-furred individual. Use a separate sheet of paper. What are the possible genotypes of the unknown parent?

Students should draw two Punnett squares. One should show a cross between two heterozygous individuals; the other should show a heterozygous individual crossed with a homozygous recessive individual. If *B* represents the dominant black fur color and *b* stands for white fur color, the unknown parent could be genotypically *Bb* or *bb*.

Quick Lab)   **CONSUMER**

Interpreting Information in a Pedigree

Teacher Notes

TIME REQUIRED 20 minutes

SKILLS ACQUIRED
Classifying
Constructing models
Recognizing patterns
Interpreting
Inferring
Organizing and analyzing information

RATING

Easy ← 1 2 3 4 → Hard

Teacher Prep–1
Student Setup–1
Concept Level–2
Cleanup–1

THE SCIENTIFIC METHOD

Analyze the Results Analysis and Conclusions question 3 requires students to analyze data presented in the form of a pedigree.

Draw Conclusions Analysis and Conclusions question 2 requires students to draw a conclusion about why pedigrees are more effective than written passages.

TIPS AND TRICKS

This lab works best in groups of two students.

Introduce the lab by reading the hereditary information presented for one of the families. Have students describe the family without looking at the pedigree. Their success should be minimal at best. Point out that the way information is organized can be crucial to using it and that this lab will show a simple way to organize hereditary information.

Discuss the mechanics of Pedigree I before students continue with Pedigree II. Extend the lab by having students construct pedigrees from summaries of novel situations, such as those involving maternal and fraternal twins, remarriage, and marriage across generations.

Provide other examples of pedigrees. Medical genetics textbooks, the genetics department of a large hospital, and agricultural breeding services are some of the sources of such pedigrees.

Name _________________________ Class ____________ Date __________

  CONSUMER

Interpreting Information in a Pedigree

Organizing information is often the key to solving a problem. Tracing the hereditary characteristics over many generations can be confusing unless the information is well organized. In this lab, you will learn how to organize hereditary information, making it much easier to analyze.

OBJECTIVES

Analyze a pedigree.

Construct a pedigree.

MATERIALS

- paper
- pencil

Procedure

1. Examine Pedigree I, which traces the dimples trait through three generations of a family. Blackened symbols represent people with dimples. Circles represent females, and squares represent males.

FIGURE 1

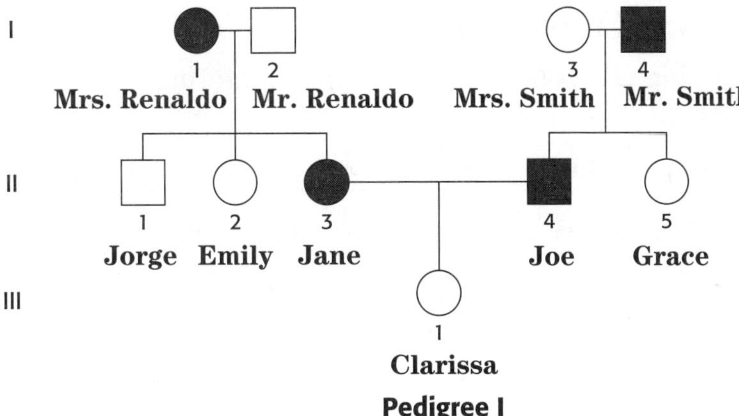

Pedigree I

2. Read the following passage, which describes the family shown in Pedigree I. Write the name of each person below the correct symbol in Pedigree I.

Although Jane and Joe Smith have dimples, their daughter, Clarissa, does not. Joe's father has dimples, but his mother, and his sister, Grace, do not. Jane's father, Mr. Renaldo, her brother, Jorge, and her sister, Emily, do not have dimples, but her mother does.

3. Look at Pedigree I again.

- How are marriage and offspring symbolized?

  A straight line connecting a circle and square indicates a marriage, with a

  descending line leading to any offspring.

Name _______________________________ Class _______________ Date _______________

Interpreting Information in a Pedigree *continued*

- What do the Roman numerals symbolize?

  Roman numerals identify each generation.

4. Construct a pedigree based on the following passage about curly hair.

Andy, Penny, and Delbert have curly hair, but their mother, Mrs. Cummins, does not. Mrs. Giordano, Mrs. Cummin's sister, has curly hair, but her parents, Mr. & Mrs. Lutz, do not. Deidra and Darlene Giordano have curly hair, but their sister, Katie, like her father, has straight hair.

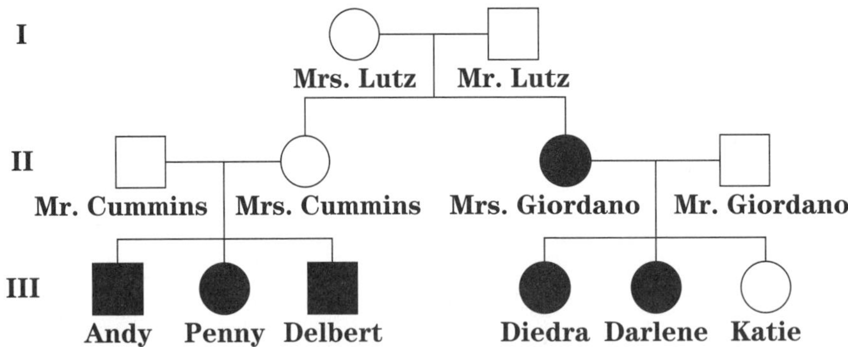

Analysis and Conclusions

1. Summarizing Observations What type of information does a pedigree contain?

A pedigree contains hereditary information, which is genetic information

about what traits are passed from one generation to the next.

2. Evaluating Models What advantages does a pedigree have over a written passage?

Answers will vary, but should indicate that a pedigree organizes hereditary

information visually, making it easier to interpret than information in a

written passage.

3. Interpreting Information Take another look at Pedigree I. A genetic counselor analyzing Pedigree I suggests that a person only needs to have one dominant allele for dimples (D) in order to have dimples. If this is true, what is the genotype of person 1 in the third generation of Pedigree I?

The genotype of person 1 in the third generation is dd. If dimples are a

dominant trait, the person would have to have two recessive alleles to not

have dimples.

Analyzing Corn Genetics

Teacher Notes

TIME REQUIRED One 45-minute class period for planting, followed by a few minutes each day for 3 to 4 weeks to water plants and collect data

SKILLS ACQUIRED

Collecting data
Experimenting
Identifying patterns
Inferring
Interpreting
Measuring
Organizing and analyzing data

RATINGS

Easy ← 1   2   3   4 → Hard

Teacher Prep–3
Student Setup–3
Concept Level–2
Cleanup–2

THE SCIENTIFIC METHOD

Make Observations Students make observations during their experiment.

Form a Hypothesis Procedure step 2 asks students to form a hypothesis.

Test the Hypothesis Procedure steps 3–5 guide students in designing an experiment that will test their hypothesis. Students conduct their experiment.

Analyze the Results Students analyze results in Analysis questions 1–3.

Draw Conclusions Conclusions question 3 asks students to draw conclusions from class data.

MATERIALS

Materials for this lab can be purchased from WARD'S. See the *Master Materials List* for ordering instructions.

SAFETY CAUTIONS

• Discuss all safety symbols with students.

• Remind students to wash their hands with antibacterial soap after the lab.

DISPOSAL

Corn plants and potting soil can be composted or reused.

Analyzing Corn Genetics *continued*

TIPS AND TRICKS

Provide enough seeds for each lab group to receive 10 seeds per lot. The plant tray or pot should be large enough to grow 30 seeds without overcrowding the seedlings.

Designate one lot of seeds (in this case, lot B) as defective (albino). Purchase albino seeds for this lot. The seeds should come in a 3:1 green to albino ratio.

Provide a container for collecting used potting soil for disposal or reuse.

Pre-Lab Discussion

Discuss with students how the seeds will be divided and distributed to each lab group. Explain that the seeds should be allowed to grow until any obvious genetic problems are observed. Ask students the following questions to guide their thinking:

- How should the corn seeds be planted? *(Seeds should be sown about 3 cm to 6 cm apart at a depth no greater than one seed's length.)*

- How often should the seeds be watered? *(Watering should be checked daily. The soil should remain damp but not drenched.)*

- What are the germination rate and survival rate, and how do they affect a farmer's crop? *(The germination rate is the percentage of planted seeds that sprout. The survival rate is the percentage of sprouted seeds that survive to maturity. Together, these two rates affect the total crop yield.)*

- How do you calculate the germination and survival rate? *(Students should use the following formulas: for germination rate,* number of germinated seeds/number of seeds planted × 100 = percent germination, *and for survival rate,* number of surviving plants/number of germinated seeds × 100 = percent survival.)

After albino corn plants appear, guide students' thinking with these questions:

- What is absent from the white (albino) corn plants? *(chlorophyll)*

- What caused the albino plants to occur? *(a mutation)*

- What will happen to the albino plants. Why? *(They will die soon after germination because, without chlorophyll, these plants are unable to produce their own food.)*

- What are the genotypes of the parents of an albino individual? *(They would have to be heterozygous because all albino, or homozygous, plants die before they can reproduce.)*

MISCONCEPTION ALERT

Make sure students understand that a Punnett square is not a predictor of crop yield. A Punnett square predicts the offspring of two individual parents. In a population of corn plants, there are many different parents with different genetic makeups. Also, environmental factors such as weather, pests, amount of sunlight, and pollution affect crop yield.

Name _______________________________ Class ______________ Date ______________

Analyzing Corn Genetics

LONG-TERM PROJECT

In this lab, you will design and conduct an experiment to compare the germination rate and survival rate of corn from normal seeds with seeds containing the mutation for albinism.

OBJECTIVES

Develop a hypothesis to explain the poor yield of a corn crop.

Design and **conduct** an experiment to test your hypothesis.

Compare germination and survival rates of three lots of corn seeds.

Evaluate your results.

POSSIBLE MATERIALS

- corn seeds (10 each from lots A, B, and C)
- lab apron
- plant tray
- safety goggles
- soil, potting (3 kg)
- water

Finding Out More Information

In simple Mendelian genetics, a physical characteristic—called a *phenotype*—is controlled by two alleles. These two alleles reside on the chromosomes and are passed to the offspring through sexual reproduction, one allele being inherited from the father and one from the mother. The two alleles are considered the organism's *genotype*, and each allele may be dominant or recessive. A *dominant* phenotype will appear if an organism has a dominant genotype—both dominant alleles, or one dominant and one recessive allele. A *recessive* phenotype will appear only if an organism has both recessive alleles.

Albinism is a characteristic of some organisms in which the cells lack pigments. Albinism is a recessive trait, so an individual with two recessive alleles will show the trait. In mammals, physical characteristics of albinism may include white hair, pale skin, and lack of coloring in the irises of the eyes. The eyes may appear pink because the blood vessels at the back of the eye become visible.

In plants, albinism is characterized by the failure to produce *chlorophyll*, a plant pigment necessary for photosynthesis. Because the trait is recessive, parents with the normal phenotype may produce seeds with the alleles for albinism. However, the seeds do not develop because they lack chlorophyll. Thus, the presence of albinism in a corn crop may prevent a farmer from being able to maximize corn production.

- What are possible genotypes of normal corn plants?

  Normal corn plants would have either both dominant alleles, or one dominant and one recessive allele.

Name _______________________________ Class _______________ Date _______________

Analyzing Corn Genetics *continued*

- How can you determine the genotype of a normal corn plant?

  One way would be to look at the offspring of the parents to determine the

  ratio of normal offspring to albino offspring. You could do this by planting

  corn seeds to see what kind of offspring result. The appearance of albino

  offspring means that both parents have the dominant and the recessive allele.

Procedure
FORMING A HYPOTHESIS

Based on what you have learned, form a hypothesis about how albinism affects corn crop yield.

1. How can a farmer ensure maximum corn production?

   A way to ensure maximum corn production is to plant corn seeds that con-

   tain only the dominant alleles so that albino plants don't show up in future

   generations.

2. Write your own hypothesis. A possible hypothesis might be "The presence of the recessive allele for albinism reduces corn yield by 25 percent."

   Answers will vary. Hypotheses should be testable in an experiment using

   materials from Possible Materials.

COMING UP WITH A PLAN

Plan and conduct an experiment that will determine the germination and survival rates of three lots of corn seeds. Be sure to control variables among the three lots. Consult with your teacher about which variables you need to consider.

3. Write out a procedure for your experiment on a separate sheet of paper. As you plan the procedure, make the following decisions.

   - Decide how far apart you will plant the seeds and how deep you will plant them.

   - Decide how often and how much you will water the seeds.

   - Decide where you will keep the seeds.

   - Decide how long you will wait for the seeds to germinate and how long after germination you will measure growth.

   - Decide what safety procedures are necessary. Add them to your written procedure.

Name _______________________________ Class _______________ Date _____________

Analyzing Corn Genetics *continued*

4. Using graph paper or a computer, construct data tables to organize your data. You might choose tables similar to those shown below. Be sure that your tables fit your investigation.

5. Have your teacher approve your plan.

PERFORMING THE EXPERIMENT

6. Put on safety goggles and a lab apron.

7. Implement your plan, using the equipment and safety procedures that you selected.

8. Record your data in your tables. If necessary, revise your tables to include variables that you did not think of while planning your experiment.

9. At the end of the experiment, pool the class data and determine the germination rate and survival rate for each lot of corn seeds.

10. When you finish, clean and store your equipment. Recycle or dispose of all materials as instructed by your teacher. Wash your hands if you handled seeds or soil.

TABLE 1 GROUP DATA

Lot	Number of seeds	Number normal plants germinated	Number albino plants germinated	Number normal plants @ 4 wk	Number albino plants @ 4 wk
A	10	9	0	9	0
B	10	7	3	7	0
C	10	8	0	8	0

TABLE 2 CLASS DATA

Lot	Number of seeds	Number normal plants germinated	Number albino plants germinated	Number normal plants @ 4 wk	Number albino plants @ 4 wk	Total germination rate (%)	Total germination rate (%)
A	50	48	0	48	0	96	100
B	50	35	12	35	0	94	74
C	50	47	0	47	0	94	100

Students' data will vary. Sample data is provided in the tables above.

Name _________________________________ Class _______________ Date ___________

Analyzing Corn Genetics *continued*

Analysis

1. Analyzing Data Using the class data, what was the germination rate for seeds in lot A? Lot B? Lot C?

Answers will vary, but will be close to 95 percent for all three lots.

2. Analyzing Data Using the class data, what was the survival rate for seeds in lot A? Lot B? Lot C? Explain any difference you observed.

Answers will vary. Lots A and C will be close to 100 percent. Lot B will be

close to 75 percent. Lot B contains albino seeds, which germinate but don't

survive due to the lack of chlorophyll.

3. Analyzing Results Did your results support your hypothesis? Explain.

Answers will vary. Students who hypothesized that albinism will reduce corn

yield or that albinism will reduce corn yield by about 25 percent will find

their hypotheses supported.

4. Recognizing Patterns Use **Figure 1** to draw a Punnett square to show how albinism is inherited in corn plants. Use G to represent the allele for chlorophyll and g to represent the allele for albinism. Explain how you developed your Punnett square.

Punnett squares should show two heterozygous parents producing a 3:1 ratio

of normal to albino offspring. Because albinism is recessive, no other combi-

nation of parents will produce albino offspring. Students can prove this by

crossing a heterozygous parent with a homozygous parent. All offspring will

be normal. Crossing two homozygous recessive individuals (albinos) is not

realistic because they would not survive to reproductive age.

Name _______________________ Class _____________ Date _____________

Analyzing Corn Genetics *continued*

FIGURE 1 PUNNETT SQUARE FOR ALBINISM IN CORN PLANTS

	G	g
G	GG	Gg
g	Gg	gg

Conclusions

1. **Evaluating Methods** Did your experimental design give clear results? If not, how might you improve your design?

   Answers will vary. If students failed to control for amount and frequency of watering, amount of light, and depth of planting among the three lots, then their results might be unclear. Controlling for these variables should yield fairly clear results.

2. **Interpreting Information** What are the genotypes of the parents of an albino individual?

   They would have to be heterozygous because all albino, or homozygous, plants die before they can reproduce.

3. **Drawing Conclusions** What does the Punnett square reveal about crop yield?

   A Punnett square predicts the expected offspring of two individual parents. It does not not predict offspring for a population, such as a corn crop.

4. **Applying Conclusions** How is the class data you and your classmates gathered in the experiment useful to a farmer planning a corn crop?

   When buying corn seed, the farmer can request that the seed come from strains of corn that have shown high germination and survival rates in lab tests, at or near 100 percent. This will help to ensure that albinos are not included in the seed that the farmer buys for planting.

Answer Key

Directed Reading

SECTION: THE ORIGINS OF GENETICS

1. Knight discovered that the purple-flowered offspring of a cross between garden peas with purple flowers and garden peas with white flowers produce some offspring with purple flowers and some with white flowers.
2. Mendel carefully counted the number of each kind of offspring, analyzed the data, and discovered that the numbers formed simple ratios.
3. Garden peas have many traits with two clearly different forms; it is easy to control matings in garden peas; and garden peas are small, mature quickly, and produce numerous offspring.
4. monohybrid
5. P
6. F_1
7. F_2

SECTION: MENDEL'S THEORY

1. f
2. a
3. g
4. b
5. c
6. e
7. d
8. y
9. tall
10. short
11. The law of segregation states that the two alleles for a trait segregate when gametes are formed during meiosis.
12. The law of independent assortment states that the pairs of alleles for different traits separate independently of one another during meiosis.

SECTION: STUDYING HEREDITY

1. recessive
2. homozygous
3. heterozygous
4. $\frac{1}{2}$
5. 0
6. $\frac{1}{2}$
7. If a trait is autosomal, it will appear equally in both sexes.
8. If an autosomal trait is dominant, every individual with the trait will have a parent with the trait. If a trait is recessive, an individual with the trait may have one, two, or no parents with the trait.

SECTION: COMPLEX PATTERNS OF HEREDITY

1. A polygenic trait is a trait that is influenced by several genes. A trait determined by multiple alleles is controlled by a gene that has three or more alleles.
2. Incomplete dominance occurs when offspring display a form of a trait that is intermediate between the forms of the trait displayed by the parents. Codominance occurs when two alleles for a characteristic are expressed at the same time.
3. proteins
4. cystic fibrosis
5. people with a family history of genetic disorders
6. Gene therapy is a procedure that attempts to replace defective genes by inserting copies of healthy genes into an individual.

Active Reading

SECTION: THE ORIGINS OF GENETICS

1. A monohybrid cross involves only one pair of contrasting traits.
2. It provides a specific example of a monohybrid cross.
3. This plant would produce only plants with white flowers.
4. The parental generation is the first two individuals that are crossed in a breeding experiment.
5. the first filial generation or first offspring of the parental generation
6. P
7. F_1
8. F_2
9. cross-pollination
10. self-pollination
11. c

SECTION: MENDEL'S THEORY

1. by writing the first letter of the trait as a capital letter
2. by writing the first letter of the trait as a lowercase letter
3. Two of the same alleles for seed color are present in the plant.
4. The plant possesses two different alleles for flower color.
5. *Yy*
6. *pp*
7. *Pp*
8. *b*

SECTION: STUDYING HEREDITY

1. It defines the key term *Punnett square*.
2. They represent the possible gametes produced by each parent.
3. Each combination is formed by taking one allele along the top of the box and one allele along the side of the box.
4. the possible genotypes of offspring produced from these two parents
5. the number of plants expressing either purple flowers or white flowers

6. *YY*	10. 2
7. *yy*	11. 1
8. 1	12. 3
9. 1	13. d

SECTION: COMPLEX PATTERNS OF HEREDITY

1. It defines the key term *multiple alleles*.
2. It clarifies the term *blood groups*, which precedes it.
3. The letters refer to two carbohydrates on the surface of red blood cells.
4. Both I^A and I^B are dominant over the recessive allele i. But neither I^A nor I^B is dominant over the other.
5. Both I^A and I^B are present in the individual. Because they are codominant, the individual shows both forms of the trait.
6. d

Vocabulary Review

1. e		9. g	
2. m		10. d	
3. j		11. l	
4. f		12. a	
5. h		13. p	
6. o		14. i	
7. k		15. b	
8. n		16. c	

17. Punnett square
18. test cross
19. probability
20. pedigree
21. sex-linked trait
22. polygenic trait
23. incomplete dominance
24. codominance
25. multiple alleles

Science Skills

ANALYZING EXPERIMENTS

1. a. *R*	i. *R*
b. *R*	j. *r*
c. *r*	k. *R*
d. *Rr*	l. *RR*
e. *Rr*	m. *Rr*
f. *r*	n. *r*
g. *Rr*	o. Rr
h. *Rr*	p. *rr*

2. Only one phenotype is present—plants with round seeds; only one genotype is present—*Rr*.
3. There are two phenotypes present—round seeds and wrinkled seeds. They are present in the ratio of three with round seeds to one with wrinkled seeds.
4. Mendel's hypothesis is supported by this analysis. When both the *R* and *r* alleles appeared together, only the dominant phenotype was expressed (round seeds).
5. Mendel probably did not observe an exact ratio of 3:1. These numbers represent probabilities. For example, there is a 75 percent chance that any one offspring will have the round phenotype. Assuming Mendel obtained large numbers of offspring to analyze, he would have calculated a ratio of approximately 3:1.

Concept Mapping

1. heredity
2. modern genetics
3. probabilities or Punnett squares
4. probabilities or Punnett squares
5. multiple alleles or mutations or polygenic traits or codominance
6. multiple alleles or mutations or polygenic traits or codominance
7. multiple alleles or mutations or polygenic traits or codominance
8. multiple alleles or mutations or polygenic traits or or codominance

Critical Thinking

1. a	13. a
2. e	14. h
3. c	15. j, e
4. d	16. l, d
5. b	17. c, f
6. f	18. k, h
7. d	19. i, b
8. b	20. g, a
9. g	21. a
10. c	22. b
11. f	23. d
12. e	24. d

Test Prep Pretest

1. d	9. a
2. c	10. b
3. d	11. a
4. b	12. d
5. d	13. c
6. a	14. b
7. c	15. b
8. c	16. b

17. Mendel
18. phenotype, genotype
19. monohybrid
20. polygenic traits
21. genetic disorders
22. Mendel calculated an approximate 3:1 ratio of contrasting traits. He derived this ratio by counting plants expressing each type of trait he was comparing. Using division, he found that the ratio of plants expressing the dominant trait to plants expressing the recessive trait was about 3:1.

23. the law of segregation and the law of independent assortment
24. In the winter, the genes of the arctic fox that code for enzymes involved in pigment production do not function because of the cold temperature. Thus, the coat of the fox is white, and the animal blends in well with its surroundings. In warmer weather, these genes function and the fox's coat darkens to a reddish brown. Fur color in Siamese cats is also influenced by temperature. The fur on the ears, nose, paws, and tail of Siamese cats is darker than the rest of their body. In plants, hydrangea flowers of the same genetic variety range in color from blue to pink, depending on the acidity of the soil. Hydrangea plants in acidic soil bloom blue flowers; those in neutral to basic soil bloom pink flowers.
25. The blood clotting disorder, hemophilia A, is a sex-linked trait. In hemophilia A, the mutation occurs on one of the genes on the X chromosome. The gene codes for a protein involved in blood clotting.

Quiz

SECTION: THE ORIGINS OF GENETICS

1. d	6. e
2. c	7. a
3. d	8. c
4. a	9. d
5. a	10. b

SECTION: MENDEL'S THEORY

1. a	6. e
2. c	7. b
3. d	8. d
4. b	9. f
5. c	10. a

SECTION: STUDYING HEREDITY

1. d	6. b
2. c	7. e
3. a	8. d
4. b	9. c
5. f	10. a

SECTION: COMPLEX PATTERNS OF HEREDITY

1. a	**6.** c
2. b	**7.** f
3. d	**8.** e
4. c	**9.** d
5. b	**10.** a

Chapter Test (General)

1. a	**11.** d
2. c	**12.** b
3. a	**13.** b
4. b	**14.** c
5. a	**15.** b
6. a	**16.** d
7. d	**17.** c
8. b	**18.** d
9. a	**19.** c
10. c	**20.** c

Chapter Test (Advanced)

1. e	**6.** g
2. j	**7.** d
3. h	**8.** f
4. a	**9.** b
5. c	**10.** i

11. F_2

12. X-linked or sex-linked

13. *Pp* and *pp*

14. 3:1

15. recessive

16. genetic counseling

17. cross-pollinate

18. self-pollinate, true-breeding

19. Some traits are controlled by several genes rather than by only one.

20. Most sex-linked characteristics are carried as alleles on the X chromosome. A male would express a sex-linked recessive trait if he gets the gene for the trait from his mother. He cannot get an X chromosome from his father. A female would have to get the gene for the recessive trait on both of the X chromosomes she receives—the one from her mother and the one from her father.

21. Examples include the changing colors of the arctic fox's fur in response to changes in the season; the darker coloration of the nose, ears, tail, and paws of a Siamese cat as a result of lower body temperature in those areas; and the different colors of hydrangea flowers resulting from different acid levels in the soil.

22. An individual with cystic fibrosis has at least one copy of a defective gene that makes a protein necessary to pump chloride into and out of cells. The airways of the lungs of these individuals become clogged with thick mucus, and the ducts of the liver and pancreas become blocked. Treatments can relieve some of the symptoms, but there is no cure for this disorder.

23. (1) For each inherited trait, an individual has two copies of the gene—one from each parent. (2) There are alternative forms of genes. Today we call them alleles. (3) When two different alleles occur together, one of them may be completely expressed (dominant), while the other may have no observable effect on the organism's appearance(recessive). (4) When gametes are formed, the alleles for each gene in an individual separate independently of one another.

24. Any four of the following: Several traits of garden peas exist in two clearly different forms, such as purple and white flowers. Male and female reproductive parts of garden peas are enclosed in the same flower, so a researcher can control mating by allowing a flower to self-fertilize or by cross-pollinating flowers. Garden peas grow easily, mature quickly, and produce many offspring—so results can be obtained quickly.

25. Mendel allowed each variety of garden pea to self-pollinate for several generations, selecting plants from each generation that exhibited one form of a trait. He did this until all of the offspring of a given variety produced only one form of a particular trait.

Lesson Plan

Section: The Origins of Genetics

Pacing

Regular Schedule: **with lab(s):** N/A   **without lab(s):** 3 days

Block Schedule: **with lab(s):** N/A   **without lab(s):** 1 1/2 days

Objectives

1. Identify the investigator whose studies formed the basis of modern genetics.

2. List characteristics that make the garden pea a good subject for genetic study.

3. Summarize the three major steps of Gregor Mendel's garden-pea experiments.

4. Relate the ratios that Mendel observed in his crosses to his data.

National Science Education Standards Covered

UNIFYING CONCEPTS AND PROCESSES

UCP1: Systems, order, and organization

UCP2: Evidence, models, and explanation

UCP3: Change, constancy, and measurement

UCP4: Evolution and equilibrium

UCP5: Form and function

SCIENCE AS INQUIRY

SI1: Abilities necessary to do scientific inquiry

SI2: Understandings about scientific inquiry

HISTORY AND NATURE OF SCIENCE

HNS1: Science as a human endeavor

HNS3: Historical perspectives

LIFE SCIENCE: MOLECULAR BASIS OF HEREDITY

LSGene1: In all organisms, the instructions for specifying the characteristics of the organisms are carried in DNA.

LSGene2: Most of the cells in a human contain two copies of each of 22 different chromosomes. In addition, there is a pair of chromosomes that determine sex.

> **KEY**
>
> **SE** = Student Edition   **TE** = Teacher Edition
> **CRF** = Chapter Resource File

Block 1

CHAPTER OPENER *(45 minutes)*

_ **Quick Review,** SE. Students answer questions covered in previous sections of the textbook as preparation for the chapter content. (**GENERAL**)

_ **Reading Activity,** SE. Students write a short list of the things that they already know about inheritance and things they would like to know about inheritance. (**GENERAL**)

_ **Using the Figure,** TE. Students answer questions about the chapter opener photograph. (**GENERAL**)

_ **Opening Activity,** TE. Students bring photographs of their parents or siblings to class and try to match the names of fellow students with the photographs by comparing similarities and differences in appearance between the identified students and their relatives. (**BASIC**)

Block 2

FOCUS *(5 minutes)*

_ **Bellringer Transparency.** Use this transparency as students enter the classroom and find their seats. (**GENERAL**)

MOTIVATE *(10 minutes)*

_ **Demonstration,** TE. Display large photographs of flowering plants. Ask students to list traits that could be inherited. (**BASIC**)

TEACH *(30 minutes)*

_ **Teaching Transparency, Section Outline.** Use this transparency to give students a framework for the information in this section. (**GENERAL**)

_ **Directed Reading Worksheet, The Origins of Genetics, CRF.** Students complete the exercises in this worksheet to help them understand the material as they read the section. (**BASIC**)

_ **Teaching Transparency, Three Steps of Mendel's Experiments.** Use this transparency to introduce Mendel's pea-plant experiments and walk students through the three main steps he followed. (**GENERAL**)

_ **Group Activity,** Benefits of Peas, TE. Students design newspaper ads that would have attracted someone like Mendel to purchase peas for genetic research. (**ADVANCED**)

HOMEWORK

_ **Active Reading Worksheet, The Origins of Genetics, CRF.** Students read a passage related to the section topic and answer questions. (**GENERAL**)

_ **Problem Solving Worksheet, Ratios and Proportions, One-Stop Planner.** Students define and give examples of ratios, rates, and proportions as they complete the practice exercises. (**GENERAL**)

Block 3

TEACH (*25 minutes*)

_ **Teaching Tip,** Hidden Traits, TE. Students try to tell which of the purple flowers shown in Figure 3 are true-breeding for purple flowers. (**GENERAL**)

_ **Math Lab,** Calculating Mendel's Ratios, TE. Students calculate ratios from data in a table and convert the ratios to fractions. (**GENERAL**)

_ **Datasheets for In-Text Labs, Calculating Mendel's Ratios, CRF.**

CLOSE (*20 minutes*)

_ **Reteaching,** TE. Groups of students make flashcards listing key terms from the lesson and then quiz each other. (**BASIC**)

_ **Quiz,** TE. Students answer questions that review the section material. (**GENERAL**)

HOMEWORK

_ **Section Review,** SE. Assign questions 1–6 for review, homework, or quiz. (**GENERAL**)

_ **Quiz, CRF.** This quiz consists of ten multiple choice and matching questions that review the section's main concepts. (**BASIC**) **Also in Spanish.**

Other Resource Options

_ **Skill Builder,** Math Skills, TE. Students practice reducing ratios to their simplest forms. (**GENERAL**)

_ **Teaching Tip,** Genetic Make-up, TE. Students estimate how many genes animals have in common with each other. (**GENERAL**)

_ **go.hrw.com.** For worksheets, videos, and other teaching aids related to this chapter, visit the HRW Web site and type in the keyword HX4 GEN.

_ **CNN Student News.** Find the latest news, lesson plans, and activities related to important scientific events at **cnnstudentnews.com.**

_ **Biology Interactive Tutor CD-ROM,** Unit 5 Heredity. Students watch animations and other visuals as the tutor explains heredity. Students assess their learning with interactive activities.

Lesson Plan

Section: Mendel's Theory

Pacing

Regular Schedule: **with lab(s):** 3 days   **without lab(s):** 2 days

Block Schedule: **with lab(s):** 1 1/2 days   **without lab(s):** 1 day

Objectives

1. Describe the four major hypotheses Mendel developed.

2. Define the terms *homozygous*, *heterozygous*, *genotype*, and *phenotype*.

3. Compare Mendel's two laws of heredity.

National Science Education Standards Covered

UNIFYING CONCEPTS AND PROCESSES

UCP1: Systems, order, and organization

UCP2: Evidence, models, and explanation

UCP4: Evolution and equilibrium

UCP5: Form and function

SCIENCE AS INQUIRY

SI1: Abilities necessary to do scientific inquiry

SI2: Understandings about scientific inquiry

HISTORY AND NATURE OF SCIENCE

HNS1: Science as a human endeavor

HNS3: Historical perspectives

LIFE SCIENCE: THE CELL

LSCell1: Cells have particular structures that underlie their functions.

LIFE SCIENCE: MOLECULAR BASIS OF HEREDITY

LSGene1: In all organisms, the instructions for specifying the characteristics of the organisms are carried in DNA.

LSGene2: Most of the cells in a human contain two copies of each of 22 different chromosomes. In addition, there is a pair of chromosomes that determine sex.

> **KEY**
> **SE** = Student Edition   **TE** = Teacher Edition
> **CRF** = Chapter Resource File

Block 4

FOCUS *(5 minutes)*

_ **Bellringer Transparency.** Use this transparency as students enter the classroom and find their seats. **(GENERAL)**

MOTIVATE *(10 minutes)*

_ **Identifying Preconceptions**, TE. Students discuss whether offspring can have traits different from those of both parents. **(BASIC)**

TEACH *(30 minutes)*

_ **Teaching Transparency, Section Outline.** Use this transparency to give students a framework for the information in this section. **(GENERAL)**

_ **Directed Reading Worksheet, Mendel's Theory, CRF.** Students complete the exercises in this worksheet to help them understand the material as they read the section. **(BASIC)**

_ **Teaching Transparency, Mendel's Factors.** Use this transparency to explain that each parent has two separate genes for a particular trait. **(GENERAL)**

_ **Teaching Tip**, Dominant and Recessive, TE. Ask students why some traits appear more often than others do and then discuss dominant and recessive. **(GENERAL)**

_ **Teaching Tip**, Genotype and Phenotype, TE. Have students practice the boldface terms in this section by providing examples for them to identify. **(GENERAL)**

HOMEWORK

_ **Active Reading Worksheet, Mendel's Theory, CRF.** Students read a passage related to the section topic and answer questions. **(GENERAL)**

Block 5

TEACH *(30 minutes)*

_ **Quick Lab**, Identifying Dominant and Recessive Traits, SE. Students determine some genotypes and all of the phenotypes for some human traits that are inherited as simple dominant or recessive traits. **(GENERAL)**

_ **Datasheets for In-Text Labs, Identifying Dominant and Recessive Traits, CRF.**

_ **Activity**, Graphic Organizer, TE. Students make a graphic organizer with a brief explanation to demonstrate the law of independent assortment. A sample graphic organizer is provided in the TE. **(GENERAL)**

_ **Activity**, Hairy Knuckles, TE. Students detemine whether they have hair above their knuckles and then relate this information to genotype. (**GENERAL**)

CLOSE *(15 minutes)*

_ **Reteaching,** TE. Students review the steps involved in scientific investigations and apply these methods to Mendel's discoveries. (**BASIC**)

_ **Alternative Assessment**, TE. Students relate Mendel's four hypotheses to his experimental results. (**GENERAL**)

_ **Quiz,** TE. Students answer questions that review the section material. (**GENERAL**)

HOMEWORK

_ **Section Review,** SE. Assign questions 1–6 for review, homework, or quiz. (**GENERAL**)

_ **Quiz, CRF.** This quiz consists of ten multiple choice and matching questions that review the section's main concepts. (**BASIC**) **Also in Spanish.**

Optional Block

LAB *(45 minutes)*

_ **Inquiry Lab, Analyzing Corn Genetics, CRF.** Students design and conduct an experiment to compare the germination rate and survival rate of corn from normal seeds with seeds containing the mutation for albinism. (**GENERAL**)

Other Resource Options

_ **Teaching Tip**, Gene Expression, TE. Students list examples of environmental influence on the expression of phenotypes. (**GENERAL**)

_ **go.hrw.com.** For worksheets, videos, and other teaching aids related to this chapter, visit the HRW Web site and type in the keyword HX4 GEN.

_ **CNN Student News.** Find the latest news, lesson plans, and activities related to important scientific events at **cnnstudentnews.com**.

_ **Biology Interactive Tutor CD-ROM,** Unit 5 Heredity. Students watch animations and other visuals as the tutor explains heredity. Students assess their learning with interactive activities.

Lesson Plan

Section: Studying Heredity

Pacing

Regular Schedule: **with lab(s):** N/A   **without lab(s):** 2 days

Block Schedule: **with lab(s):** N/A   **without lab(s):** 1 day

Objectives

1. Predict the results of monohybrid genetic crosses by using Punnett squares.

2. Apply a test cross to determine the genotype of an organism with a dominant phenotype.

3. Predict the results of monohybrid genetic crosses by using probabilities.

4. Analyze a simple pedigree.

National Science Education Standards Covered

UNIFYING CONCEPTS AND PROCESSES

UCP1: Systems, order, and organization

UCP2: Evidence, models, and explanation

UCP3: Change, constancy, and measurement

UCP4: Evolution and equilibrium

UCP5: Form and function

SCIENCE AS INQUIRY

SI1: Abilities necessary to do scientific inquiry

SI2: Understandings about scientific inquiry

HISTORY AND NATURE OF SCIENCE

HNS1: Science as a human endeavor

HNS3: Historical perspectives

LIFE SCIENCE: MOLECULAR BASIS OF HEREDITY

LSGene1: In all organisms, the instructions for specifying the characteristics of the organisms are carried in DNA.

LSGene2: Most of the cells in a human contain two copies of each of 22 different chromosomes. In addition, there is a pair of chromosomes that determine sex.

> **KEY**
> **SE** = Student Edition **TE** = Teacher Edition
> **CRF** = Chapter Resource File

Block 6

FOCUS *(5 minutes)*

_ **Bellringer Transparency.** Use this transparency as students enter the classroom and find their seats. (**GENERAL**)

MOTIVATE *(10 minutes)*

_ **Discussion/Question**, TE. Students hypothosize a genetic explanation for why basenji dogs cannot bark. (**GENERAL**)

TEACH *(30 minutes)*

_ **Teaching Transparency, Section Outline.** Use this transparency to give students a framework for the information in this section. (**GENERAL**)

_ **Teaching Transparency, Monohybrid Crosses of Homozygous Plants and Heterozygous Plants.** Use this transparency to compare the two crosses shown. Discuss monohybrid crosses and review the terms *monohybrid, dominant, recessive, homozygous,* and *heterozygous.* (**GENERAL**)

_ **Using the Figure**, Figure 8, TE. Use this strategy to discuss the parts of a Punnett square. Then have students practice using a Punnett square for several monohybrid crosses. (**BASIC**)

_ **Teaching Transparency, Probability with Two Coins.** Use this transparency to discuss the concept of probability using a coin toss. (**GENERAL**)

_ **Data Lab**, Analyzing a Test Cross, SE. Students use a Punnett square to do a test cross to determine genotypes. (**GENERAL**)

_ **Datasheets for In-Text Labs, Analyzing a Test Cross, CRF.**

HOMEWORK

_ **Directed Reading Worksheet, Studying Heredity, CRF.** Students complete the exercises in this worksheet to help them understand the material as they read the section. (**BASIC**)

_ **Active Reading Worksheet, Studying Heredity, CRF.** Students read a passage related to the section topic and answer questions. (**GENERAL**)

_ **Problem Solving Worksheet, Genetics and Probability, CRF.** Students prepare simple crosses and calculate the possible offspring that could result. Students determine the probability that each type of offspring will occur. (**GENERAL**)

Block 7

TEACH *(35 minutes)*

_ **Demonstration**, TE. Students determine the probability of drawing an ace from a deck of shuffled cards. **(GENERAL)**

_ **Math Lab**, Predicting the Results of Crosses Using Probabilities, SE. Students practice using probabilities to predict the outcome of genetic crosses. **(GENERAL)**

_ **Datasheets for In-Text Labs, Predicting the Results of Crosses Using Probabilities, CRF.**

_ **Teaching Tip**, Sex Linked, TE. Students discuss how a male might inherit a sex-linked trait from his mother. **(BASIC)**

_ **Data Lab**, Evaluating a Pedigree, SE. Students practice interpreting a pedigree. They determine whether the trait is sex linked or autosomal and whether it is dominant or recessive. **(GENERAL)**

_ **Datasheets for In-Text Labs, Evaluating a Pedigree, CRF.**

CLOSE *(10 minutes)*

_ **Reteaching,** TE. Students work in pairs and to construct and compare Punnett squares showing a cross between two genotypes. **(BASIC)**

_ **Quiz,** TE. Students answer questions that review the section material. **(GENERAL)**

HOMEWORK

_ **Section Review,** SE. Assign questions 1–5 for review, homework, or quiz. **(GENERAL)**

_ **Quiz, CRF.** This quiz consists of ten multiple choice and matching questions that review the section's main concepts. **(BASIC) Also in Spanish.**

Other Resource Options

_ **Exploring Further**, Crosses That Involve Two Traits, SE. Students complete and analyze a dihybrid cross. **(GENERAL)**

_ **Teaching Tip**, Using Probabilities in Genetic Crosses, TE. Point out that the Punnett square can be used to determine the probability that a specific   genotype will occur. **(GENERAL)**

_ **Internet Connect.** Students can research Internet sources about Genetic Counseling with SciLinks Code HX4090.

_ **Internet Connect.** Students can research Internet sources about Genetic Disorders with SciLinks Code HX4091.

_ **go.hrw.com.** For worksheets, videos, and other teaching aids related to this chapter, visit the HRW Web site and type in the keyword HX4 GEN.

- **CNN Student News.** Find the latest news, lesson plans, and activities related to important scientific events at **cnnstudentnews.com**.

- **Biology Interactive Tutor CD-ROM,** Unit 5 Heredity. Students watch animations and other visuals as the tutor explains heredity. Students assess their learning with interactive activities.

- **CNN Science in the News, Video Segment 7 Gene Progress.** This video segment is accompanied by a **Critical Thinking Worksheet**.

Lesson Plan

Section: Complex Patterns of Heredity

Pacing

Regular Schedule: **with lab(s):** 3 days   **without lab(s):** 2 days

Block Schedule: **with lab(s):** 1 1/2 days   **without lab(s):** 1 day

Objectives

1. Identify the factors that influence patterns of heredity.

2. Describe how mutations can cause genetic disorders.

3. List two genetic disorders, and describe their causes and symptoms.

4. Evaluate the benefits of genetic counseling.

National Science Education Standards Covered

UNIFYING CONCEPTS AND PROCESSES

UCP1: Systems, order, and organization

UCP2: Evidence, models, and explanation

UCP3: Change, constancy, and measurement

UCP4: Evolution and equilibrium

UCP5: Form and function

SCIENCE AS INQUIRY

SI1: Abilities necessary to do scientific inquiry

SI2: Understandings about scientific inquiry

HISTORY AND NATURE OF SCIENCE

HNS1: Science as a human endeavor

HNS3: Historical perspectives

LIFE SCIENCE: THE CELL

LSCell1: Cells have particular structures that underlie their functions.

LSCell2: Most cell functions involve chemical reactions.

LIFE SCIENCE: MOLECULAR BASIS OF HEREDITY

LSGene1: In all organisms, the instructions for specifying the characteristics of the organisms are carried in DNA.

LSGene2: Most of the cells in a human contain two copies of each of 22 different chromosomes. In addition, there is a pair of chromosomes that determine sex.

> **KEY**
> **SE** = Student Edition **TE** = Teacher Edition
> **CRF** = Chapter Resource File

Block 8

FOCUS *(5 minutes)*

_ **Bellringer Transparency.** Use this transparency as students enter the classroom and find their seats. **(GENERAL)**

MOTIVATE *(10 minutes)*

_ **Discussion/Question**, TE. Students propose a mechanism for the inheritance of a trait such as eye color in humans. **(GENERAL)**

TEACH *(30 minutes)*

_ **Teaching Transparency, Section Outline.** Use this transparency to give students a framework for the information in this section. **(GENERAL)**

_ **Integrating Physics and Chemistry**, TE. Students identify and describe the limitations of Mendel's understanding of inheritance based on his pea plant experiments. **(GENERAL)**

_ **Teaching Tip**, Incomplete Dominance, TE. Students discuss whether a   plant breeder could produce only pink-flowering snapdragons by crossing pink flowering and white-flowering snapdragons. **(BASIC)**

_ **Demonstration**, TE. Set up four flasks as listed on page   176 to demonstrate universal donor and acceptor concepts. Take an empty beaker,   and pour "O" into it. Show the students that pouring "A" into the beaker of "O" would contaminate the "O." Demonstrate the possible mixtures. **(GENERAL)**

HOMEWORK

_ **Directed Reading Worksheet, Complex Patterns of Heredity, CRF.** Students complete the exercises in this worksheet to help them understand the material as they read the section. **(BASIC)**

_ **Active Reading Worksheet, Complex Patterns of Heredity, CRF.** Students read a passage related to the section topic and answer questions. **(GENERAL)**

Block 9

TEACH *(35 minutes)*

_ **Integrating Physics and Chemistry**, TE. Students conduct an experiment in the lab in order to demonstrate how pH affects the everyday physical expression of traits. **(GENERAL)**

- **Teaching Tip**, Hemoglobin, TE. Students discuss how a faulty gene can alter a hemoglobin molecule. (**ADVANCED**)

- **Teaching Transparency, Some Human Genetic Disorders.** Use this transparency to summarize important information about some common   genetic disorders. (**GENERAL**)

- **Group Activity**, Patterns of Heredity, TE. Students make a table to organize information about patterns of heredity that are more complex than simple dominant-recessive patterns. (**GENERAL**)

CLOSE *(10 minutes)*

- **Reteaching**, K-W-L, TE. Students return to their lists of things they want to learn about inheritance, which they created at the beginning of this chapter. Students should check off the questions that they can now answer and make a list of what they have learned. Review any questions they still have. (**BASIC**)

- **Quiz**, TE. Students answer questions that review the section material. (**GENERAL**)

HOMEWORK

- **Alternative Assessment**, TE. Students create design an informative brochure about an assigned genetic disorder. (**GENERAL**)

- **Section Review**, SE. Assign questions 1–6 for review, homework, or quiz. (**GENERAL**)

- **Quick Lab, Interpreting Information in a Pedigree, CRF.** Students construct and analyze a pedigree. (**BASIC**)

- **Quiz, CRF.** This quiz consists of ten multiple choice and matching questions that review the section's main concepts. (**BASIC**) **Also in Spanish.**

- **Modified Worksheet, One-Stop Planner.** This worksheet has been specially modified to reach struggling students. (**BASIC**)

- **Science Skills Worksheet, CRF.** Students analyze a breeding experiment using Punnett squares. (**GENERAL**)

- **Critical Thinking Worksheet, CRF.** Students answer analogy-based questions that review the section's main concepts and vocabulary. (**ADVANCED**)

Optional Block

LAB *(45 minutes)*

- **Skills Practice Lab,** Modeling Monohybrid Crosses, SE. Students simulate a monohybrid cross and then calculate genotypic and phenotypic ratios. (**GENERAL**)

- **Datasheets for In-Text Labs, Modeling Monohybrid Crosses, CRF.**

Other Resource Options

- **Supplemental Reading, The Double Helix, One-Stop Planner.** Students read the book and answer questions. (**ADVANCED**)

- **Reading Skill Builder**, Reading Organizer, TE. Students students make a reading organizer describing the causes and effects of each of the genetic disorders discussed in this section. (**BASIC**)

- **Internet Connect.** Students can research Internet sources about Genetic Counseling with SciLinks Code HX4090.

- **Internet Connect.** Students can research Internet sources about Genetic Disorders with SciLinks Code HX4091.

- **go.hrw.com.** For worksheets, videos, and other teaching aids related to this chapter, visit the HRW Web site and type in the keyword HX4 GEN.

- **CNN Student News.** Find the latest news, lesson plans, and activities related to important scientific events at **cnnstudentnews.com**.

- **Biology Interactive Tutor CD-ROM,** Unit 5 Heredity. Students watch animations and other visuals as the tutor explains heredity. Students assess their learning with interactive activities.

- **CNN Science in the News, Video Segment 7 Gene Progress.** This video segment is accompanied by a **Critical Thinking Worksheet**.

Lesson Plan

End-of-Chapter Review and Assessment

Pacing

Regular Schedule: 2 days

Block Schedule: 1 day

KEY

SE = Student Edition   **TE** = Teacher Edition

CRF = Chapter Resource File

Block 10

REVIEW *(45 minutes)*

- **Study Zone,** SE. Use the Study Zone to review the Key Concepts and Key Terms of the chapter and prepare students for the Performance Zone questions. **(GENERAL)**

- **Performance Zone,** SE. Assign questions to review the material for this chapter. Use the assignment guide to customize review for sections covered. **(GENERAL)**

- **Teaching Transparency, Concept Mapping.** Use this transparency to review the concept map for this chapter. **(GENERAL)**

Block 11

ASSESSMENT *(45 minutes)*

- **Chapter Test, Mendel and Heredity, CRF.** This test contains 20 multiple choice and matching questions keyed to the chapter's objectives. **(GENERAL) Also in Spanish.**

- **Chapter Test, Mendel and Heredity, CRF.** This test contains 25 questions of various formats, each keyed to the chapter's objectives. **(ADVANCED)**

- **Modified Chapter Test, One-Stop Planner.** This test has been specially modified to reach struggling students. **(BASIC)**

Other Resource Options

- **Vocabulary Review Worksheet, CRF.** Use this worksheet to review the chapter vocabulary. **(GENERAL) Also in Spanish.**

- **Test Prep Pretest, CRF.** Use this pretest to review the main content of the chapter. Each question is keyed to a section objective. **(GENERAL) Also in Spanish.**

- **Test Item Listing for ExamView® Test Generator, CRF.** Use the Test Item Listing to identify questions to use in a customized homework, quiz, or test.

- **ExamView® Test Generator, One-Stop Planner.** Create a customized homework, quiz, or test using the HRW Test Generator program.

Mendel and Heredity

TRUE/FALSE

1. ____ Genetics is the branch of biology that involves the study of how different traits are transmitted from one generation to the next.
   Answer: True   Difficulty: I   Section: 1   Objective: 1

2. ____ Mendel discovered predictable patterns in the inheritance of traits.
   Answer: True   Difficulty: I   Section: 1   Objective: 1

3. ____ Garden peas are difficult to grow because they mature slowly.
   Answer: False   Difficulty: I   Section: 1   Objective: 2

4. ____ The mating of garden-pea flowers can be easily controlled because the male and female reproductive parts are enclosed within the same flower.
   Answer: True   Difficulty: I   Section: 1   Objective: 2

5. ____ When Mendel cross-pollinated two varieties from the P generation that exhibited contrasting traits, he called the offspring the second filial, or F_2, generation.
   Answer: False   Difficulty: I   Section: 1   Objective: 3

6. ____ Mendel's initial experiments were monohybrid crosses.
   Answer: True   Difficulty: I   Section: 1   Objective: 3

7. ____ In Mendel's experiments, the recessive traits appeared in the F_2 generation in approximately 25 percent of the plants.
   Answer: True   Difficulty: I   Section: 1   Objective: 4

8. ____ A dominant allele masks the effect of a recessive allele.
   Answer: True   Difficulty: I   Section: 2   Objective: 1

9. ____ Individuals must exhibit a trait in order for it to appear in their offspring.
   Answer: False   Difficulty: I   Section: 2   Objective: 1

10. ____ The allele for a recessive trait is usually represented by a capital letter.
    Answer: False   Difficulty: I   Section: 2   Objective: 2

11. ____ Heterozygous individuals have two of the same alleles for a particular gene.
    Answer: False   Difficulty: I   Section: 2   Objective: 2

12. ____ In heterozygous individuals, only the recessive allele is expressed.
    Answer: False   Difficulty: I   Section: 2   Objective: 2

13. ____ The law of segregation states that two or more pairs of alleles separate independently of one another during gamete formation.
    Answer: False   Difficulty: I   Section: 2   Objective: 3

14. ____ A Punnett square represents the phenotype of an organism.
    Answer: False   Difficulty: I   Section: 3   Objective: 1

15. ____ If the offspring of a test cross all have the dominant trait, then the genotype of the individual being tested is homozygous.
    Answer: True   Difficulty: I   Section: 3   Objective: 2

16. ____ Probability is the likelihood that a certain event will occur.
    Answer: True   Difficulty: I   Section: 3   Objective: 3

17. ____ The expression of sex-linked genes is controlled by hormones.
    Answer: False   Difficulty: I   Section: 3   Objective: 4

18. ____ An autosomal trait will occur with equal frequency in both males and females.
    Answer: True   Difficulty: I   Section: 3   Objective: 4

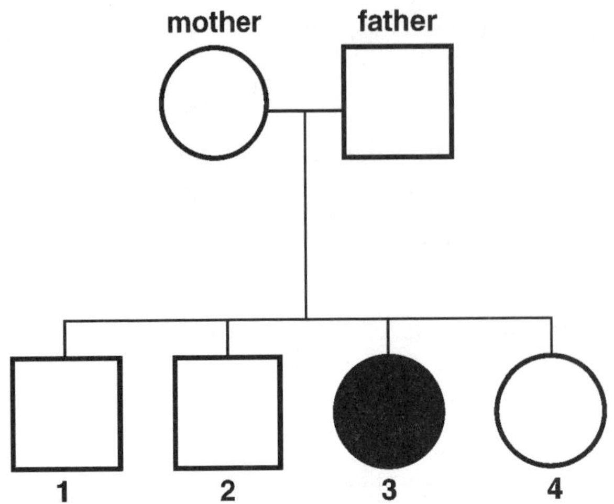

19. ____ Refer to the illustration above. The father listed in the pedigree is most likely heterozygous for the trait.
    Answer: True   Difficulty: II   Section: 3   Objective: 4

20. ____ Refer to the illustration above. Child #3 probably has a homozygous recessive phenotype.
    Answer: True   Difficulty: II   Section: 3   Objective: 4

21. ____ Refer to the illustration above. The trait indicated in the pedigree is sex-linked.
    Answer: False   Difficulty: II   Section: 3   Objective: 4

22. ____ In codominance, two dominant alleles are expressed at the same time.
    Answer: True   Difficulty: I   Section: 4   Objective: 1

23. ____ All genes have only two alleles.
    Answer: False   Difficulty: I   Section: 4   Objective: 1

24. ____ The only way a mutation in a recessive gene can show up in a child born to two normal parents is for both parents to be heterozygous.
    Answer: True   Difficulty: I   Section: 4   Objective: 2

25. ____ Hemophilia is caused by a mutated allele that produces a defective form of the protein hemoglobin.
    Answer: False   Difficulty: I   Section: 4   Objective: 3

26. ____ Cystic fibrosis is a genetic disorder caused by a defective chloride-ion transport protein.
    Answer: True   Difficulty: I   Section: 4   Objective: 3

27. ____ Genetic counselors often help people with a family history of genetic disorders.
    Answer: True   Difficulty: I   Section: 4   Objective: 4

MULTIPLE CHOICE

28. The passing of traits from parents to offspring is called
    a. genetics.
    b. heredity.
    c. development.
    d. maturation.
    Answer: B   Difficulty: I   Section: 1   Objective: 1

29. The difference between Mendel's experiments in the area of heredity and those done by earlier researchers was that
    a. earlier researchers did not have microscopes.
    b. earlier researchers used detailed and numerical procedures.
    c. Mendel expressed the results of his experiments in terms of numbers.
    d. Mendel used pea plants with both purple and white flowers.

    Answer: C   Difficulty: I   Section: 1   Objective: 1

30. The scientific study of heredity is called
    a. meiosis.
    b. crossing-over.
    c. genetics.
    d. pollination.

    Answer: C   Difficulty: I   Section: 1   Objective: 1

31. The "father" of genetics was
    a. T.A. Knight.
    b. Hans Krebs.
    c. Gregor Mendel.
    d. None of the above

    Answer: C   Difficulty: I   Section: 1   Objective: 1

32. Mendel obtained his P generation by allowing the plants to
    a. self-pollinate.
    b. cross-pollinate.
    c. assort independently.
    d. segregate.

    Answer: A   Difficulty: I   Section: 1   Objective: 3

33. Step 1 of Mendel's garden pea experiment, allowing each variety of garden pea to self-pollinate for several generations, produced the
    a. F_1 generation.
    b. F_2 generation.
    c. P generation.
    d. P_2 generation.

    Answer: A   Difficulty: I   Section: 1   Objective: 3

34. $F_2 : F_1 ::$
    a. $P : F_1$
    b. $F_1 : F_2$
    c. $F_1 : P$
    d. dominant trait : recessive trait

    Answer: C   Difficulty: II   Section: 1   Objective: 3

35. Mendel's law of segregation states that
    a. pairs of alleles are dependent on one another when separation occurs during gamete formation.
    b. pairs of alleles separate independently of one another after gamete formation.
    c. each pair of alleles remains together when gametes are formed.
    d. the two alleles for a trait separate when gametes are formed.

    Answer: D   Difficulty: I   Section: 2   Objective: 3

36. Garden peas
    a. are difficult to grow.
    b. mature quickly.
    c. produce few offspring.
    d. are not good subjects for studying heredity.

    Answer: B   Difficulty: I   Section: 1   Objective: 2

37. The phenotype of an organism
    a. represents its genetic composition.
    b. is the physical appearance of a trait.
    c. occurs only in dominant pure organisms.
    d. cannot be seen.

    Answer: B   Difficulty: I   Section: 2   Objective: 2

38. If an individual possesses two recessive alleles for the same trait, the individual is said to be
    a. homozygous for the trait.
    b. haploid for the trait.
    c. heterozygous for the trait.
    d. mutated.

    Answer: A   Difficulty: I   Section: 2   Objective: 2

39. A genetic trait that appears in every generation of offspring is called
    a. dominant.
    b. phenotypic.
    c. recessive.
    d. superior.

    Answer: A   Difficulty: I   Section: 2   Objective: 2

40. An individual heterozygous for a trait and an individual homozygous recessive for the trait are crossed and produce many offspring that are
    a. all the same genotype.
    b. of two different phenotypes.
    c. of three different phenotypes.
    d. all the same phenotype.

    Answer: B   Difficulty: I   Section: 2   Objective: 2

41. Tallness (T) is dominant to shortness (t) in pea plants. Which of the following represents a genotype of a pea plant that is heterozygous for tallness?
    a. T
    b. TT
    c. Tt
    d. tt

    Answer: C   Difficulty: I   Section: 2   Objective: 2

42. homozygous : heterozygous ::
    a. heterozygous : Bb
    b. probability : predicting chances
    c. dominant : recessive
    d. factor : gene

    Answer: C   Difficulty: II   Section: 2   Objective: 2

43. Mendel's finding that the inheritance of one trait had no effect on the inheritance of another became known as the
    a. law of dominance.
    b. law of universal inheritance.
    c. law of separate convenience.
    d. law of independent assortment.

    Answer: D   Difficulty: I   Section: 2   Objective: 3

44. The discovery of chromosomes provided a link between the first law of heredity that stemmed from Mendel's work and
    a. pollination.
    b. inheritance.
    c. mitosis.
    d. meiosis.

    Answer: D   Difficulty: I   Section: 2   Objective: 3

45. A 3:1 ratio of tall to short pea plants appearing in the F_2 generation lends support to the law of
    a. recessiveness.
    b. mutation.
    c. segregation.
    d. crossing-over.

    Answer: C   Difficulty: I   Section: 2   Objective: 3

46. The law of segregation states that
    a. alleles of a gene separate from each other during meiosis.
    b. different alleles of a gene can never be found in the same organism.
    c. each gene of an organism ends up in a different gamete.
    d. each gene is found on a different molecule of DNA.

    Answer: A   Difficulty: I   Section: 2   Objective: 3

In humans, having freckles (*F*) is dominant to not having freckles (*f*). The inheritance of these traits can be studied using a Punnett square similar to the one shown below.

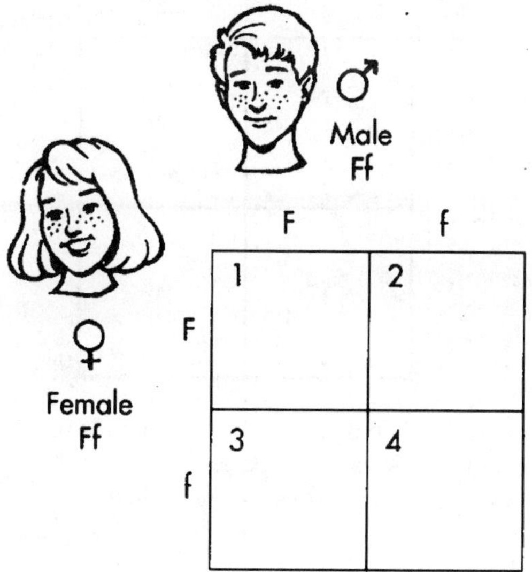

47. Refer to the illustration above. The child represented in box 1 in the Punnett square would
    a. be homozygous for freckles.
    b. have an extra freckles chromosome.
    c. be heterozygous for freckles.
    d. not have freckles.

    Answer: A   Difficulty: II   Section: 3   Objective: 1

48. Refer to the illustration above. The parents shown in the Punnett square could have children with a phenotype ratio of
    a. 1:2:1.
    b. 4:0.
    c. 3:1.
    d. 2:2.

    Answer: C   Difficulty: II   Section: 3   Objective: 1

49. Refer to the illustration above. Which box in the Punnett square represents a child who does *not* have freckles?
    a. box 1
    b. box 2
    c. box 3
    d. box 4

    Answer: D   Difficulty: II   Section: 3   Objective: 1

50. Refer to the illustration above. The child in box 3 of the Punnett square has the genotype
    a. *FF*.
    b. *Ff*.
    c. *ff*.
    d. None of the above

    Answer: B   Difficulty: II   Section: 3   Objective: 1

In rabbits, black fur (*B*) is dominant to brown fur (*b*). Consider the following cross between two rabbits.

Bb x Bb

	B	**b**
B	1	2
b	3	4

51. Refer to the illustration above. The device shown, which is used to determine the probable outcome of genetic crosses, is called a
    a. Mendelian box.
    b. Punnett square.
    c. genetic graph.
    d. phenotypic paradox.

    Answer: B   Difficulty: II   Section: 3   Objective: 1

52. Refer to the illustration above. Both of the parents in the cross are
    a. black.
    b. brown.
    c. homozygous dominant.
    d. homozygous recessive.

    Answer: A   Difficulty: II   Section: 3   Objective: 1

53. Refer to the illustration above. The phenotype of the offspring indicated by box 3 would be
    a. brown.
    b. black.
    c. a mixture of brown and black.
    d. None of the above

    Answer: B   Difficulty: II   Section: 3   Objective: 1

54. Refer to the illustration above. The genotypic ratio of the F_1 generation would be
    a. 1:1.
    b. 3:1.
    c. 1:3.
    d. 1:2:1.

    Answer: D   Difficulty: II   Section: 3   Objective: 1

55. What is the expected genotypic ratio resulting from a homozygous dominant $\times$ heterozygous monohybrid cross?
    a. 1:0
    b. 1:1
    c. 1:2:1
    d. 1:3:1

    Answer: B   Difficulty: II   Section: 3   Objective: 1

56. What is the expected genotypic ratio resulting from a heterozygous $\times$ heterozygous monohybrid cross?
    a. 1:2:1
    b. 1:3:1
    c. 1:2
    d. 1:0

    Answer: A   Difficulty: II   Section: 3   Objective: 1

57. What is the expected phenotypic ratio resulting from a homozygous dominant $\times$ heterozygous monohybrid cross?
    a. 1:3:1
    b. 1:2:1
    c. 2:1
    d. 1:0

    Answer: D   Difficulty: II   Section: 3   Objective: 1

58. The unknown genotype of an individual with a dominant phenotype can be determined using
   a. a ratio.
   b. a dihybrid cross.
   c. probability
   d. a test cross.

   Answer: D   Difficulty: I   Section: 3   Objective: 2

59. What is the probability that the offspring of a homozygous dominant individual and a homozygous recessive individual will exhibit the dominant phenotype?
   a. 0.25
   b. 0.5
   c. 0.66
   d. 1.0

   Answer: D   Difficulty: II   Section: 3   Objective: 3

60. A trait that occurs in 450 individuals out of a total of 1,800 individuals occurs with a probability of
   a. 0.04.
   b. 0.25.
   c. 0.50.
   d. 0.75.

   Answer: B   Difficulty: II   Section: 3   Objective: 3

61. If a characteristic is sex-linked, it
   a. occurs most commonly in males.
   b. occurs only in females.
   c. can never occur in females.
   d. is always fatal.

   Answer: A   Difficulty: I   Section: 3   Objective: 4

62. Since the allele for colorblindness is located on the X chromosome, colorblindness
   a. cannot be inherited.
   b. occurs only in adults.
   c. is sex-linked.
   d. None of the above

   Answer: C   Difficulty: I   Section: 3   Objective: 4

63. A diagram in which several generations of a family and the occurrence of certain genetic characteristics are shown is called a
   a. Punnett square.
   b. monohybrid cross.
   c. pedigree.
   d. family karyotype.

   Answer: C   Difficulty: I   Section: 3   Objective: 4

64. In humans, the risks of passing on a genetic disorder to offspring can be assessed by
   a. analysis of a pedigree.
   b. genetic counseling.
   c. prenatal testing.
   d. All of the above

   Answer: D   Difficulty: I   Section: 3   Objective: 4

65. How many different phenotypes can be produced by a pair of codominant alleles?
   a. 1
   b. 2
   c. 3
   d. 4

   Answer: C   Difficulty: II   Section: 4   Objective: 1

66. Which of the following traits is controlled by multiple alleles in humans?
   a. sickle cell anemia
   b. blood type
   c. hemophilia
   d. Huntington's disease

   Answer: B   Difficulty: I   Section: 4   Objective: 1

67. What would be the blood type of a person who inherited an *A* allele from one parent and an *O* allele from the other?
   a. type A
   b. type B
   c. type AB
   d. type O

   Answer: A   Difficulty: II   Section: 4   Objective: 1

68. A change in a gene due to damage or being copied incorrectly is called
   a. evolution.
   b. meiosis.
   c. segregation.
   d. a mutation.

   Answer: D   Difficulty: I   Section: 4   Objective: 2

69. Which of the following describes hemophilia?
    a. multiple-allele trait
    b. dominant trait
    c. sex-linked trait
    d. codominant trait

    Answer: C   Difficulty: I   Section: 4   Objective: 3

70. Both sickle-cell anemia and hemophilia
    a. are caused by genes coding for defective protein.
    b. are seen in homozygous dominant individuals.
    c. provide resistance to malaria infections.
    d. are extremely common throughout the world.

    Answer: A   Difficulty: I   Section: 4   Objective: 3

71. Genetic counseling is a process that
    a. helps identify parents at risk for having children with genetic defects.
    b. assists parents in deciding whether or not to have children.
    c. uses a family pedigree.
    d. All of the above

    Answer: D   Difficulty: I   Section: 4   Objective: 4

72. Which of the following is an example of gene technology?
    a. A genetic counselor studies a pedigree.
    b. A student studies the colors of flowers in pea plants.
    c. A geneticist explains the inheritance of albinism using a Punnett square.
    d. A physician transfers a normal gene into the DNA of a person with a genetic disease.

    Answer: D   Difficulty: II   Section: 4   Objective: 4

COMPLETION

73. The patterns that Mendel discovered form the basis of ___________________, the branch
    of biology that deals with heredity.

    Answer: genetics   Difficulty: I   Section: 1   Objective: 1

74. The passing of traits from parents to offspring is called ___________________.

    Answer: heredity   Difficulty: I   Section: 1   Objective: 1

75. A reproductive process in which fertilization occurs within a single plant is
    ___________________ ___________________.

    Answer: self-pollination   Difficulty: I   Section: 1   Objective: 2

76. The transferring of pollen between plants is called ___________________-
    ___________________.

    Answer: cross-pollination   Difficulty: I   Section: 1   Objective: 2

77. A(n) ___________________ cross is a cross that involves one pair of contrasting traits.

    Answer: monohybrid   Difficulty: I   Section: 1   Objective: 3

78. Mendel called the offspring of the P generation the ___________________ generation.

    Answer: F_1 Difficulty: I   Section: 1   Objective: 3

79. In Mendel's experiments, a trait that disappeared in the F_1 generation but reappeared in
    the F_2 generation was always a(n) ___________________ trait.

    Answer: recessive   Difficulty: II   Section: 1   Objective: 4

80. A trait that is not expressed in the F_1 generation resulting from the crossbreeding of two
    genetically different, true-breeding organisms is called ___________________.

    Answer: recessive   Difficulty: I   Section: 2   Objective: 1

81. Different forms of a particular gene are called ___________________.

    Answer: alleles   Difficulty: I   Section: 2   Objective: 1

82. In heterozygous individuals, only the __________________ allele is expressed.
    Answer: dominant   Difficulty: I   Section: 2   Objective: 2

83. An organism that has two identical alleles for a trait is called __________________.
    Answer: homozygous Difficulty: I   Section: 2   Objective: 2

84. An organism's __________________ refers to the set of alleles it has inherited.
    Answer: genotype   Difficulty: I   Section: 2   Objective: 2

85. The external appearance of an organism as determined by what alleles are present is
    __________________.
    Answer: phenotype   Difficulty: I   Section: 2   Objective: 2

86. The statement that the members of each pair of alleles separate when gametes are formed
    is known as the law of __________________.
    Answer: segregation Difficulty: I   Section: 2   Objective: 3

87. The principle that states that alleles of different genes separate independently of one
    another during gamete formation is the law of __________________
    __________________.
    Answer: independent assortment Difficulty: I Section: 2   Objective: 3

In pea plants, tallness (*T*) is dominant to shortness (*t*). Crosses between plants with these
traits can be analyzed using a Punnett square similar to the one shown below.

	T	t
T	1	2
t	3	4

88. Refer to the illustration above. The parents shown in the Punnett square could have
    offspring with a genotypic ratio of __________________.
    Answer: 1:2:1   Difficulty: II   Section: 3   Objective: 1

89. Refer to the illustration above. Box 2 and box __________________ in the Punnett
    square represent plants that would be heterozygous for the trait for tallness.
    Answer: 3   Difficulty: II   Section: 3   Objective: 1

90. Refer to the illustration above. The phenotype of the plant that would be represented in
    box 4 of the Punnett square would be __________________.
    Answer: short   Difficulty: II   Section: 3   Objective: 1

91. Refer to the illustration above. The genotype of both parents shown in the Punnett square
    above is __________________.
    Answer: *Tt*   Difficulty: II   Section: 3   Objective: 1

92. If some of the offspring of a test cross have the recessive trait, then the genotype of the
    individual being tested is __________________.
    Answer: heterozygous Difficulty: I   Section: 3   Objective: 1

93. The likelihood that a specific event will occur is called __________________.
    Answer: probability   Difficulty: I   Section: 3   Objective: 3

94. A trait that is determined by a gene that is only found on the X chromosome is said to be
    __________________-__________________.
    Answer: sex-linked   Difficulty: I   Section: 3   Objective: 4

95. Identifying patterns of inheritance within a family over several generations is possible by studying a diagram called a(n) ___________________.

    Answer: pedigree   Difficulty: I   Section: 3   Objective: 4

96. A situation in which two dominant alleles are expressed at the same time is called

    ___________________.

    Answer: codominance Difficulty: I   Section: 4   Objective: 1

97. A trait controlled by three or more alleles is said to have ___________________

    ___________________.

    Answer: multiple alleles Difficulty: I   Section: 4   Objective: 1

98. A phenomenon in which a heterozygous individual has a phenotype that is intermediate between the phenotypes of its two homozygous parents is called ___________________

    ___________________.

    Answer: incomplete dominance Difficulty: I   Section: 4   Objective: 1

99. A change in an organism's DNA is called a(n) ___________________.

    Answer: mutation   Difficulty: I   Section: 4   Objective: 2

100. A person who is heterozygous for a recessive disorder is called a(n)

    ___________________.

    Answer: carrier   Difficulty: I   Section: 4   Objective: 2

101. A genetic disorder resulting in defective blood clotting is ___________________.

    Answer: hemophilia Difficulty: I   Section: 4   Objective: 3

102. Fragile blood cells with an irregular shape that may block blood vessels is a symptom of a genetic disease known as ___________________ ___________________

    ___________________.

    Answer: sickle cell anemia Difficulty: I   Section: 4   Objective: 3

103. It may be possible to cure genetic disorders through the use of ___________________ technology.

    Answer: gene   Difficulty: I   Section: 4   Objective: 4

PROBLEM

104. In tomato plants, tallness is dominant over dwarfness, and hairy stems are dominant over hairless stems. True-breeding (homozygous) plants that are tall and have hairy stems are available. True-breeding (homozygous) plants that are dwarf and have hairless stems are also available. Design an experiment to determine whether the genes for height and hairiness of the stem are on the same or different chromosomes. Explain how you will be able to determine from the results whether the genes are on the same or different chromosomes.

    Answer:

    The experiment should be designed to produce F_1 plants that are then allowed to pollinate each others' flowers and produce an F_2 generation of plants. If the F_2 generation has four different phenotypes present in approximate proportions of 9/16 tall and hairy, 3/16 tall and hairless, 3/16 dwarf and hairy, and 1/16 dwarf and hairless, then the student can conclude that the genes for height and hairiness are on different chromosomes. If the F_2 generation has only two different phenotypes present in approximate proportions of 3/4 tall and hairy and 1/4 dwarf and hairless, then the student can conclude that the genes for height and hairiness are on the same chromosome. The student could also conclude that the genes are located very close to each other on the chromosome. If the F_2 generation has four different phenotypes with the tall and hairless types composing less than 3/16 of the total number and the dwarf and hairy types composing less than 3/16 of the total number, then the student could

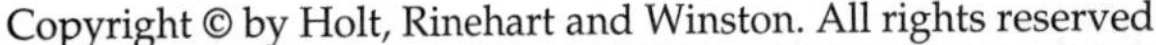

conclude that the genes for height and hairiness are on the same chromosome but not located adjacent to each other.

Difficulty: III   Section: 3   Objective: 1

ESSAY

105. Briefly discuss the reasons that Mendel chose the pea plant, *Pisum sativum*, as the organism to study in his experiments.

Answer:

The pea plant, *Pisum sativum*, is an ideal organism for genetic studies for several reasons. There are a number of traits that are easily identified and tracked from generation to generation. Each of these traits has two forms, one of which regularly disappears and reappears in alternate generations. Also, this species is easy to grow and matures quickly. Finally, gametes of both sexes are found in the same flower, so cross-pollination is easy to accomplish by removing the anthers from some flowers and transferring pollen from others to the remaining pistils.

Difficulty: III   Section: 1   Objective: 2

106. Describe pollination in pea plants.

Answer:

The reproductive structures of seed plants are located inside the flowers. In pea plants, each flower has both male and female structures. The male reproductive parts, the anthers, produce pollen grains that contain sperm. The female reproductive structure produces the egg. The tip of the female structure is called the stigma. Pollination is the transfer of pollen from anthers to stigma.

Difficulty: III   Section: 1   Objective: 2

107. In what ways did Mendel's methods help ensure his success in unraveling the mechanics of heredity?

Answer:

Mendel's choice of plants to study was fortunate since pea plants displayed several traits in contrasting forms. His use of large numbers of samples allowed the gathering of statistically significant amounts of data. In addition, he kept very careful records and used logical, orderly methods that minimized the possibility of errors.

Difficulty: III   Section: 1   Objective: 3

108. Describe how genotype and phenotype are related.

Answer:

The genetic makeup of an organism is its genotype. The external appearance of an organism is its phenotype. The phenotype of an organism is determined to a large degree by the genotype of the organism. Environmental factors and other factors can influence the phenotype of an organism.

Difficulty: III   Section: 2   Objective: 2

109. Explain what is meant by homozygous and heterozygous.

Answer:

When both alleles of a pair are the same, an organism is said to be homozygous for that characteristic. An organism may be homozygous dominant or homozygous recessive. A pea plant that is homozygous dominant for height will have the genotype *TT*. A pea plant that is homozygous recessive for height will have the genotype *tt*. When the two alleles in the pair are not the same—for example, when the genotype is *Tt*—the organism is heterozygous for that characteristic.

Difficulty: III   Section: 2   Objective: 2

110. What hypotheses did Gregor Mendel develop based on his observations of pea plants?

Answer:
1. For each inherited trait, an individual has two copies of the gene—one from each parent.
2. There are alternative versions of genes (which Mendel called factors).
3. When two different alleles occur together, one of them may be completely expressed, while the other may have no observable effect on the organism's appearance.
4. When gametes are formed, the alleles for each gene in an individual separate independently of one another. Thus, gametes carry only one allele for each inherited trait. When gametes unite during fertilization, each gamete contributes one allele.

Difficulty: III   Section: 2   Objective: 1

111. Describe Mendel's principle of independent assortment.

Answer:
From his work on pea plants, Mendel concluded that factors for different characteristics are not connected. He stated the principle of independent assortment: Factors for different characteristics are distributed to reproductive cells independently.

Difficulty: III   Section: 2   Objective: 3

112. What are three ways to express the probability of an event that occurs 500 times out of 2,000 total trials?

Answer:

The general formula for probability is $\frac{\text{number of one kind of event}}{\text{number of all events}}$

This may be expressed as a ratio ($\frac{500}{2,000}$, or $\frac{1}{4}$), as a decimal (0.25), or as a percentage (25 percent).

Difficulty: III   Section: 3   Objective: 3

113. In humans, colorblindness is a recessive, sex-linked trait. What is the likelihood that the children of a woman heterozygous for colorblindness and a man with normal color vision will be colorblind? Explain your answer.

Answer:
Since all the female offspring receive the normal allele for vision from the father, all female offspring will have normal color vision, although half of them will receive the recessive allele from the mother and thus be carriers. Since all of the male offspring receive the Y chromosome from the father, it is the X chromosome they receive from the mother that will determine whether or not they are colorblind. Since the mother is heterozygous, male offspring will have a 50 percent chance of being colorblind.

Difficulty: III   Section: 3   Objective: 4

114. Discuss the inheritance pattern that would be seen in a pedigree designed to study a recessive sex-linked characteristic.

Answer:
Sex-linked characteristics are carried on alleles on the X chromosome. As a result, sex-linked recessive traits are rarely seen in a female, unless she is the offspring of an affected male and a female who is a carrier or is affected. Males born to a female who is either a carrier or affected may inherit the sex-linked allele. If the female is affected, both of her X chromosomes will possess the gene under study, and the male is sure to inherit it. If she is a carrier, only one of her X chromosomes will possess the sex-linked allele, and the male will have a 50-50 chance of inheriting this gene. Since the male has only one X chromosome, he will not have a dominant allele in his genotype to counteract the effect of the sex-linked allele.

Difficulty: III   Section: 3   Objective: 4

115. Describe what is meant by multiple alleles, and give an example.

Answer:

Some traits are determined by more than two alleles. When three or more alleles control a trait, it is said to have multiple alleles. For example, the trait of blood type in humans is determined by multiple alleles.

Difficulty: III   Section: 4   Objective: 1

116. All of the offspring resulting from a cross between a red snapdragon and a white snapdragon are pink. What is the possible explanation for this?

Answer:

Incomplete dominance is the phenomenon that occurs when two or more alleles influence a phenotype. In other words, the offspring displays a trait that is intermediate to a trait exhibited by each parent.

Difficulty: III   Section: 4   Objective: 1